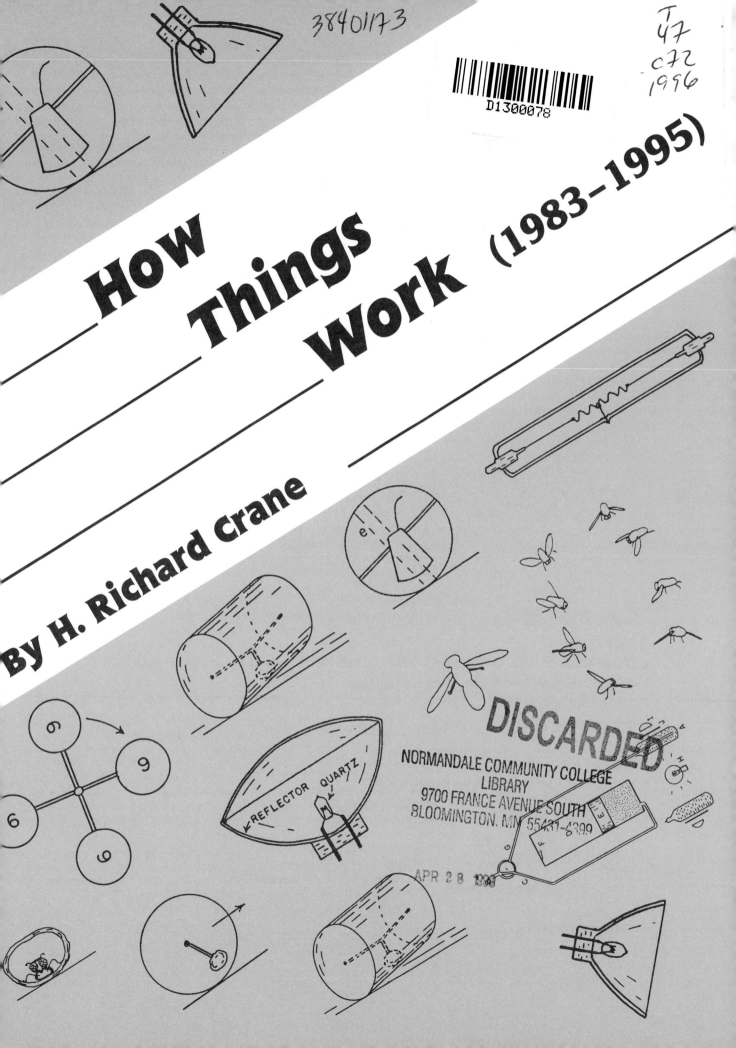

How Things Work (1983–1995)

By H. Richard Crane

How Things Work
(A Collection of "How Things Work" Columns from *The Physics Teacher*, 1983–1995.)
©1996 American Association of Physics Teachers
Second Printing, 1994
Third Printing, 1996

Published by:
   American Association of Physics Teachers
   One Physics Ellipse
   College Park, MD 20740-3845
   U.S.A.

Cover:
   The cover was designed by Rebecca Heller Rados.

ISBN #: 0-917853-44-X

Contents

1992

1993

1994

1995

Preface

Since 1983, Dick Crane has been telling readers of *The Physics Teacher* (*TPT*) "How Things Work." The topics have ranged from toys to spider webs, and from ring interferometers to gasoline pumps. All the devices and phenomena are ones that we meet in everyday life, all involve physics principles, and all require explanations that are not immediately obvious. Among his many other talents, Dick Crane not only figures out the inner workings of these things, but also makes the explanations understandable to us. During the past 12 years, *TPT* has carried over 80 examples of this talent.

The "How Things Work" column in *TPT* is just one out of a long list of contributions that Dick Crane has made to physics and physics teaching. His exquisite precision measurement of the g-factor of the electron was recognized in 1967 by the award of the Davisson-Germer prize by The American Physical Society. He is a member of the National Academy of Sciences, a Fellow of the American Academy of Arts and Sciences, and winner of a National Medal of Science in 1986. During his long career at the University of Michigan at Ann Arbor, he served as chairman of the Physics Department for seven years and as director of a defense project for the University during World War II. Within our own association of AAPT, he is a past president and has been awarded the Oersted Medal and the Melba Newell Phillips Award. He has served as president of the Midwestern Universities Research Association and as chairman of the Governing Board of the American Institute of Physics. Although retired from formal teaching since 1978, Dick continues many science and education projects besides his work for *TPT*. As a major activity he contributes his expertise to the Ann Arbor Hands-On Museum. He has created a number of unique exhibits for the Museum, as well as having written a guide book for visitors that explains the underlying science in the things they see and work with. In all this his flair for experimental devices and simple explanation shows through.

Why should we now collect all these columns about "How Things Work" into a single book? They are available in past issues of *The Physics Teacher*, and Dick promises that there will be no end to future unpuzzlings. New mysteries of everyday life continually make their appearance, providing more grist for Dick's mill. On the other hand, there is a large audience of young physics teachers who may not have complete sets of *TPT*. Furthermore, all of us will appreciate the convenience of having the topics gathered and organized. Be sure to read Dick's explanation of how he writes these columns, printed on page 78 of this volume. Then you must decide whether to leave the book on your coffee table for casual browsing or on your office reference shelf. Perhaps you should get two copies, or three if you are foolish enough to loan a book like this.

Cliff Swartz

Cliff Swartz
Editor, *The Physics Teacher*

Light Dimmer

I found this described in a book, but I had to do "reverse engineering" on a real one (take it apart, trace the circuit and measure the components) to satisfy myself. The heart of it is a triac, which is two silicon controlled rectifiers (SCR's) in parallel but with polarities opposite, in a single chip (Fig. 1a and 1b). An SCR is triggered into its ON (anode to cathode conducting) state when a small positive voltage is applied to the gate. Once ON, the gate has no further control; the SCR will continue ON until the anode-cathode current is interrupted or reduced nearly to zero. In the dimmer a triac is used instead of an SCR, because current has to be controlled in both directions. To reduce the average current, the triac is triggered to the conducting state late in each half cycle of the 60-Hz ac. The lagging gate voltage is obtained from a resistance-capacitance phase shifter, R and C_2, as shown in Fig. 1c. After a given SCR of the triac is triggered, late, it conducts only until the line voltage next crosses zero, so current flows through the lamp for only part of that half cycle. The phase lag of the gate can be varied (by the resistance) from zero to nearly a full half cycle, so the light can be dimmed almost to zero. The other components, L and C_1, serve to prevent electrical "hash" from going back over the power line. The triac goes into conduction suddenly, so L and C_1 are needed to "round off" the sharp current changes. The phase shifter in the specimen I took apart had R = 0 – 500 000 Ω, and C_2 = 0.05 μF.

The triac dimmer is remarkable. My specimen is the smallest size, yet it is rated at 600 W, and it dissipates only one watt per ampere. Contrast this with dimming a light by connecting a resistance in series. There you would be running an electric heater as well as a light! Or think of a variable transformer such as a variac. For 600 W, that would weigh 10 lb or more and cost a small fortune. Triac controls are used for fans, drills, etc., but one should not use a light dimmer for those purposes. Such loads are

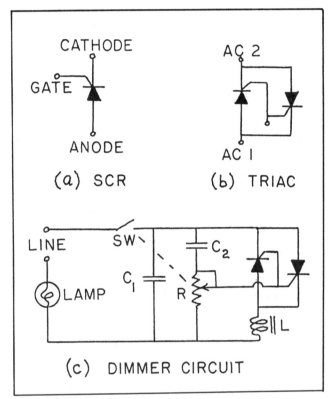

Fig. 1. Schematic

inductive, and they require a somewhat different control circuit.

Interesting questions arise. Does the electric meter charge honestly for a dimmed light, when the current is far different from a sine wave? What if *all* lights and other loads in a city were reduced by triacs? The load on the generating station would be absent for the first part of every half cycle — which surely would make a problem.

Frisbees, Can Lids, and Gyroscopic Effects

As a kid you probably sailed flat stones and tin can lids. They didn't stay horizontal; they slowly turned toward, or into, the vertical plane and fell to earth more or less edgewise. The frisbee stays horizontal, and that is an important reason it is such an ingenious invention. Let's see where the difference lies. Because the tin can lid (or other disk) is spinning, we suspect right away that the

turning is gyroscopic precession. But for that to happen there must be a torque. So we look at the two forces acting on the disk, those of gravity and of the air, to see if together they give a torque. Imagine we are riding with the disk and can see the air flowing by, above it and below it (Fig. 1a). The disk is shown inclined just a little to its direction of motion, so it will have lift. It will have the character of a flat "wing," which is well understood. Most of the deflection of the air occurs ahead of the center, so the resulting upward force acts on a line that passes ahead of the center of the disk (and center of gravity) as indicated by the arrow. That force and the gravity force together give a torque around an axis in the disk that runs in and out of the paper, as we look at the figure. That is all that is needed to make the flying disk act like a precessing gyroscope. The relationship among the axes of spin, y, the applied torque, z and the precession, x, true for any gyroscope, are shown in Fig. 1b. As related to our flying disk, x is also the approximate direction of flight. So as seen by the thrower the disk turns slowly (Fig. 1c), counterclockwise if the spin is in the sense shown.

Consider now the frisbee. The one thing certain is that torque is absent, for there is no precession. The net force of the air must act on a line very nearly through the center of gravity. And that condition evidently holds over a large range of velocity, for even as the frisbee slows almost to a stop, precession does not seem to appear. That has been accomplished through the special contour of the frisbee. It is an "airfoil," and as such is in the realm of empirical design, best done in a wind tunnel. With only a little imagination, the air flow around the frisbee and the net forces can be sketched, as in Fig. 1d.

Besides staying flat, the frisbee sails remarkably far, and that suggests that the airfoil shape gives it a very favorable lift-to-drag ratio (a term that is self-explanatory, the drag being the backward force that slows the frisbee down). We asked an aeronautical engineer about the foregoing. He answered only by telling how to make some tests! We should find the "glide angle." Sail a frisbee off the top of a building at such a downward angle that the velocity stays constant. Then the path is a straight line and there is no acceleration, so the force analysis is simple. Doing the comparison test with a flat disk might not be so easy, particularly because of precession. But the questions might interest students on a fine summer day, since they will be throwing frisbees anyway. We would like to hear of any results.

Related devices come to mind. The "clay pidgeons" used in trap shooting are shaped much like the frisbee. They come out of the mechanical launcher spinning, convex side up, and fly straight and far. Is there more than a chance relation? Consider the discus, surely designed without much help from physicists. Because of the large mass and angular momentum of the flying discus the gyro effects and the lift and drag due to the air are not expected

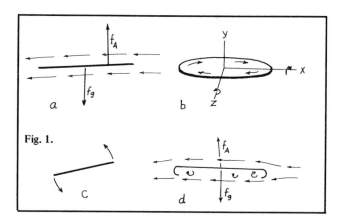

Fig. 1.

to have much effect on the trajectory. Nevertheless, no effect is too small to win or lose in the competition. If allowed, could our knowledge about frisbees lead to greater discus records? Another example is the boomerang. There precession plays an important (and somewhat complicated) role in making the object return to the thrower. The physics and aerodynamics of the boomerang are now quite well understood.[1] That they can even be thrown successfully by persons other than tribesmen is suggested by the fact that they are a commercial item in this country.[2] The effect of the frisbee on concepts of UFO's is left to the reader!

References

1. "The Amateur Scientist" department of *The Scientific American*, March and April, 1979.
2. Edmund Scientific Co. 1982 Fall Catalog.

More questions (Have you sent us yours?)

Ultrasonic small-animal and insect repellers. Recently these have been advertised widely. What is the transducer, the frequency, power, etc.? Can dogs hear them? Cats?

Instant fluorescent light starter. First, how is the non-instant kind started, then what is the change that makes it instant? Will the same tubes work on both?

Automatic focusing in cameras. Within recent years this has been offered in several makes of cameras. Are there several systems? Can the physics of it be explained simply?

Wrist watches that include a readout of pulse rate. They have appeared in catalogs recently. How is the pulse picked up, and how is it sorted from the noise?

Touch Panels in Elevators and Idiosyncrasies of Gas Tubes

Physicists often while away the time of an elevator trip speculating as to what is behind the little square that lights up when it is touched. But the trip is over before the answer is found. The part that is touched is solid: it is not a mechanical push-button. Does it depend on conduction through the finger? Or is it capacitive coupling? It works through a leather glove, but not a mitten — unless you have come in out of the rain. Only one guess seems safe: the color indicates a neon discharge.

Having heard that many years ago Peter Franken[1] had taken the direct approach and disassembled a live touch panel, I tried the same. But everything I did called the elevator — sometimes with people. So I quit. Rescue came when Dave Shalda, the University's expert on elevators, supplied a circuit diagram and a spare "glow tube" with which to experiment.

The glow tube proved to be just a modified version of a neon lamp. The characteristics of it that are essential for touch operation are present also in the garden variety neon lamp, and that makes it easy to study and demonstrate touch operation with materials readily at hand. I connected up the circuit of Fig. 1a, using a variable dc power supply. The lamp was an NE45, Fig. 1b. The resistor shown in the diagram is the one that is inside the metal base of the lamp. (Most neon lamps with screw or bayonet bases have built-in resistors.) As the voltage of the power supply was turned up, the discharge did not "strike" until about 80 V was reached. Once started, it did not go out until the voltage was turned back to about 60. Thus in the interval between 60 and 80 V, the discharge is capable of continuing *if started*, but it will not start by itself, and that is the first requirement for making a touch switch possible.

The second, and only other, requirement is that there be a simple way to trigger the discharge, when the voltage is set within the interval 60 to 80 V. This happened very reliably with the NE45 when the glass bulb was touched, the area of greatest sensitivity being that nearest the anode. (The anode is the piece that stays dark; the cathode the one that becomes sheathed in the orange glow.) I tried a few other things. With ac instead of dc voltage the discharge could be triggered, but could be kept going only by holding the finger on the glass. That's because it goes out and has to be restarted every time the voltage passes through zero, 120 times a second. The familiar "wheat seed" neon lamp, Fig. 1c, could be triggered, but not as sensitively as the NE45. An external resistance of 20 000 Ω was used. *Caution*: The above makes a good demonstration, but fingers will wander! Make sure the high-voltage metal parts and wires are taped.

Moving from here to controlling an elevator is engineering. The current is routed through the coil of a relay, which starts the chain of events that calls or stops the elevator. Switches activated by the elevator interrupt the circuit to extinguish the discharge when the elevator arrives. The special design of the gas tube increases the sensitivity, the light output, and the interval between the spontaneous striking and extinguishing voltages. The specimen at hand is sketched in Fig. 1d. The anode, *l*, is a wire that extends to within a millimeter or two of the inner surface of the glass. It is shielded, except at the tip, by a glass tube, *m*. The cathode, *n*, over which the orange glow resides, has a large area, annular in shape. On the outside of the glass there is a spot of a transparent conducting film, *o*. (There is a third electrode, a grid, consisting of a ring of wire just above the cathode, not used in

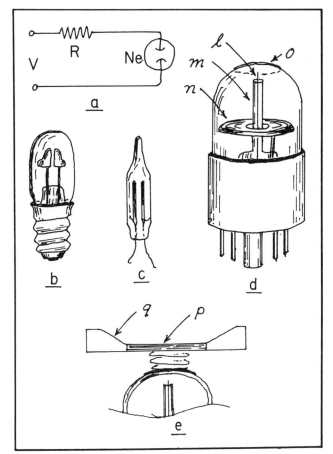

Fig. 1. All sketches are to the same scale: the glass bulb in d is 1 in. in diameter.

systems having only one elevator, and not shown here.) In operation the supply voltage is 135, which is midway between the spontaneous striking and extinguishing voltages. The external resistance is such that the current is about 30 mA; enough to give a bright glow.

Of course the elevator rider does not touch the gas tube itself. The inner square, *p*, in Fig. 1e is of conducting material, and is connected to the conducting patch on the outside of the gas tube by a coiled wire. The surrounding frame, *q*, is translucent plastic, through which the orange light comes.

Unanswered questions remain. How does touching the outside disturb the electric field inside enough to start the discharge? Is it the capacitance-to-ground of the body? The static electricity we always carry around? Or the 60-Hz ac we pick up by being antennas? If an elevator is handy, experiment. Touch the panel with a fine wire that is held in the hand, or grounded, or connected through a small capacitor — or a resistor. Stick thin plastic on the touch area. But don't leave it there!

More questions (Send yours today.)

Air purifiers for rooms.[2] What do they remove and how? Light pens that work on a computer's CRT.[3] How does the computer know what the pen is pointing at? What if it is pointing at a dark area?

Non-mechanical phonograph or video disk pickup. No record wear.
Cordless telephone extensions.

Notes

1. Chm., Optical Sciences Center of University of Arizona

2. Sent by Glen L. Green, physics teacher, Mundelein High School, Mundelein, IL 60060.
3. Sent by John S. Wallingford, Pembroke State University, Pembroke, NC 28372.

Liquid Crystal Displays: Watches, Calculators, and (Soon) Cars

For most of us the mention of liquid crystals calls to mind the display on a watch. But that association is recent. Liquid crystals have fascinated experimenters for nearly a century. There are several classes of them, each class having its distinct kind of ordered structure, and each showing a variety of types of behavior.[1-4] One of the classes of liquid crystal, in a special kind of display "sandwich" is commonly used in watches and calculators, and may soon become familiar in automobile instrument panels. We have a limited objective here: to describe the workings of the sandwich that is used in the above applications. Most of what will follow was communicated by Dr. George W. Smith of the General Motors Research Laboratories, where, as one might guess, the practical interest in liquid crystal displays (LCD's) points toward the instrumentation of cars. The Corvette has already been so converted.

In simple terms, the LCD works on light-polarization effects, at the heart of which is the remarkable ability of a liquid crystal to rotate the plane of polarization of light on command by a small electric voltage. A layer of the liquid, sandwiched between pieces of the familiar Polaroid film as polarizer and analyzer, forms a controllable light valve, to stop or pass light, and therefore to make a given area appear light or dark.

We follow Dr. Smith's sketch of the sandwich, Fig. 1a. The liquid crystal is confined between glass plates G. Next are Polaroid films P and A, in crossed orientation. Finally, a reflecting surface R. Transparent, electrically conducting films, E are applied on the glass plates. Postponing for the moment the matter of forming numbers or letters, how does this configuration turn the light off and on? In the liquid crystal the long thin molecules prefer to align themselves more or less parallel to one another. At the boundary, which is the coated glass plate, they align themselves with any "grain" that exists in the surface. That grain is provided simply by rubbing the surface, before the sandwich is assembled. The rubbing makes micro-grooves, and they are sufficient to cause the molecules to align parallel to them at the boundary. The directions of rubbing of the two facing surfaces are at right angles. To conform to the rub-directions at the two boundaries, and also be more or less parallel to one another throughout, the direction of the molecules executes a gradual 90° twist, in the distance between the plates (Fig. 1a). Now comes the even more surprising fact: the plane of polarization of the light follows the twist, turning 90° in the short distance of a hundredth of a millimeter,

which is typical of the separation of the glass plates. The light therefore can traverse the crossed Polaroid films and emerge (L). The area looks light.

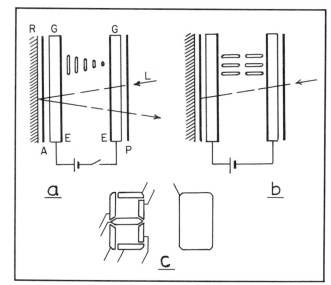

Fig. 1. Configuration of elements in liquid crystal display.

The molecules of the liquid crystal are active not only optically, but electrically. The application of an electric field by a small voltage between the plates orients them parallel to the field, perpendicular to the plates (Fig. 1b). The twist is gone, so the light is stopped at the second Polaroid film. The area looks dark. The field strength required is considerable: while only a single-cell battery is used (in a watch) the small separation of the plates results in a field of over 1000 V per cm.

All that remains is to see how the numbers and letters are made. The transparent conducting film on one plate is etched to form 7-segment characters, with a connection out from each segment to the solid-state chip (Fig. 1c). The opposing plate has only a rectangle, which covers the segments, but not the connections.

There are variations on the configuration described. If the Polaroid films have parallel, not crossed, pass directions, the characters will be light on dark background. Crossed Polaroids give dark characters on light background. The reflector can have a color filter on it. For viewing in darkness as well as in daylight, the rear coating can be made half-reflecting, with a light source (white or colored) located behind it. When applied to a driving or flying machine, the LCD will be backed up by a microprocessor chip, so that calculated information, like estimated time of arrival, as well as primary data like engine temperature, will be displayed. A given LCD can give various messages, in automatic rotation or at the call of the operator; the limit will lie only in the variety of sensors that can be planted in the machine to feed data in. (The Corvette panel reads out 14 items, in English or metric units, in 9 display areas, in numbers and graphs, in color.) Who will need the car radio any more for entertainment?

More questions

Kite tails. When we were kids kites had to have long tails to keep them from nose diving. Few tails are seen now. What's the secret?

Dolby noise suppression (in record players, and recently in theaters).

References

1. Andrew P. Penz, "Hydro-Optic Effects in Liquid Crystals," Phys. Teach. **13**, 199 (1975).
2. George H. Heilmeier, "Liquid crystal display devices," Sci. Amer. April 1970, p. 100.
3. James L. Fergason, "Liquid crystals," Sci. Amer. August 1964, p. 77.
4. Anthony J. Nicastro, "Demonstrations of some optical properties of liquid crystals," Phys. Teach. **21**, 181 (1983).

How the Housefly Uses Physics to Stabilize Flight

In conducting a column of this sort, the temptation is to report on the amazing gadgets invented in the last decade or so, mostly based on integrated circuit chips. For a change of pace I would like to discourse on a clever physics device that comes from millions of years ago — that which the housefly relies upon to maintain equilibrium in flight. A larger scale example of it came to the attention of M. L. Foucault in the mid-1800s, when he was making parts for a clock that was to be used for guiding a telescope. He was working on a thin steel rod, held in the chuck of a lathe. He noticed that if the rod was "twanged," set into vibration in a plane, and the chuck was then rotated, the plane of vibration did not follow the rotation of the chuck, but remained fixed in the laboratory coordinates. That set him to thinking along lines that led him to the invention of the Foucault pendulum. (He says so in a footnote to his first article on the pendulum.[1]) From there he followed the same line of thinking and invented the gyroscope. But back to the housefly. The fly uses, for continuously sensing changes in his (excuse me for omitting his/her) orientation in space, a version of the twanged rod.

Flies belong to the order Diptera, the term signifying that they have only one pair of wings. In place of the hind pair of wings that other flying insects have, they have a pair of *halteres* — stiff stalks, loaded at their ends by knobs, as shown in Fig. 1a. The stalks are hinged at their bases, and during flight they oscillate in arcs of about 50°, at a frequency of several hundred hertz. The hinge is such that the motion is constrained to a plane that is fixed with respect to the fly. (The plane of motion, the arc, and other features are similar enough to those of the wings to suggest strongly that the halteres evolved from wings.) When the fly changes orientation during flight, the halteres, like Foucault's twanged rod, would "like" to maintain their planes of vibration in the laboratory frame. But because of the constraint of the hinge, the planes must change with the fly, and that results in torques at the hinges. The torques are sensed by nerves, and those signals are used by the fly to execute maneuvers, or to correct orientation if he is trying to fly straight. How do we know all of this?

As early as the beginning 1700s experiments were reported in which it was found that if the halteres were cut off, the fly became disoriented in flight. In the same era it was found that if, on a fly without halteres, a thread was attached to the back end of the abdomen, like a kite tail, the fly could fly straight — interesting, but explaining little about the halteres! An understanding of the physics of the problem did not come until well into the present century — more than 200 years later. (It seems unlikely that Foucault knew of the problem, for if it had come to his attention after his observations on the twanged rod in about 1850, he probably would have seen the solution.) In the present century many hypotheses were put forward as to the action of the halteres, leading nowhere, and not until about 1938 was the dynamic (called by workers in the field "gyroscopic") action hit upon. In 1948 a monumental piece of

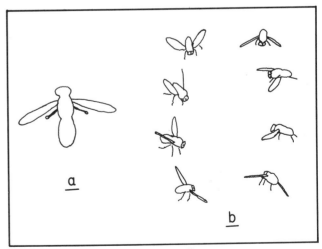

Fig. 1. **a.** Location of the halteres on a fly.
    b. Flight path of a fly without halteres, showing disorientation in yaw, sketched from the figure in Pringle's article. The sequence is counterclockwise from the top left, 1/8 s from each position to the next.

work was published by J. W. S. Pringle,[2] which seems to have answered most of the questions.

I will recount a few of the highlights of what Pringle found, but reading the original (37 pp) article is recommended. A fly was fastened to a platform that could be rotated, and electrodes were placed in the various nerves at the base of the halteres. While the fly's wings and halteres were buzzing and the platform was rotating, the torques on the halteres caused certain nerves to "fire" during *each stroke* so that the signal was a series of pulses at the vibration frequency (or twice that) rather than an average of the torque. The most effective signals were produced by yaw (rotation about the fly's vertical axis). That finding was supported by a rapid series of flash photos of a fly whose halteres had been removed. The disorientation was seen to be mainly in yaw. I have attempted to sketch the result, from the figure in Pringle's paper (Fig. 1b). Pringle says that (as of 1948) he could not find a mathematical analysis of the dynamics (he calls it gyroscopic effect) of an oscillating rod (and presumably not of the twanged rod either), so he presents the analysis in his paper. That, taking into account the known planes of vibration of the halteres, shows that yaw is sensed more strongly than either of the other rotations, pitch and roll. Pringle proposes a mechanical model made of a rod, hinge, springs, etc. One of the features in it (which he found in the fly) is that the hinge is elastic, so the motion is at the resonant frequency with a moderately high Q, kept going by a muscle pulse at one (not both) end of each stroke. Pringle remarks that, with one or two dubious exceptions, there are no other examples in the animal kingdom in which the gyroscopic effect is used.

The experimenter may feel the urge to try something. Flies are still expendable.

References and notes

1. Jour. Franklin Inst. Ser 3, Vol 21, pp. 350-53, 1851. (English translation from Comptes Rendus Acad. Sci. Paris)
2. J. W. S. Pringle, Phil. Transactions, Royal Soc. London, B, Vol 233, pp. 347-84, 1948.

More questions (I need your questions. I need information about questions asked in this and previous columns. You do not have to write an article — I'll do the work and look for further information if necessary, if you can give me a good start on an explanation.)

Touch-tone control — phones, model airplanes, garage door openers.

Digital sound recording. Can it be done simply enough for this column?

More about the frisbee. David G. Stork of Swarthmore College writes to me "as one bird to another" to explain a gyroscopic effect in frisbee sailing that I did not mention in my recent column,[1] and which in fact I had not seen. It is the maneuver of skipping. As he describes it, if the frisbee is thrown at just the right glancing angle toward a smooth horizontal surface (such as blacktop), it will touch briefly, skip up, and continue in flight. He says it is quite easy to skip a frisbee in that way — the trick lies in insuring that the proper edge of the frisbee touches the blacktop, and that is where a knowledge of gyroscopic motion comes in.

We recall that in the earlier column we identified the three principal axes as the line of flight, the axis of spin, and the transverse axis which is horizontal and perpendicular to the line of flight. At that time we were concerned with precession about the line of flight, as caused by a torque about the transverse axis. But for the skip, we need precession about the transverse axis, so the frisbee's leading edge will tilt up and cause it to "take off" and fly again. That requires a torque about the line of flight, and in the correct sense, so the nose will go up, not down. David Stork tells how to accomplish that. If you are right handed and throw in the standard way, the spin will be clockwise (viewed from above). You must throw it a little tilted, so that the left side (as you view it) will make contact with the blacktop. Then the clockwise torque around the direction of flight, from the force of the blacktop will give precession that will tilt the leading edge up. Stork shows this with a diagram of angular momentum vectors, which you also can do easily. Next time I see an expert sailing a frisbee, I'll ask for a demonstration.

Reference

1. Phys. Teach. **21**, 325 (1983).

Making Light Bulbs Last Forever
or, "There is no such thing as a free lunch"

This started when Albert Bartlett of the University of Colorado, and former president of our AAPT, agreed to look into the "bulb savers" (devices for extending the life by lowering the filament temperature) that are now widely advertised and available in hardware stores. He enlisted the help of Gary Geissinger of the Aerospace Systems Division of the Ball Corporation, who made measurements on one

kind of saver. That got me started measuring, on a saver of a different kind.

Why the interest in savers anyway? Simply that there are many applications in which greater light-bulb life, at the sacrifice of some efficiency in converting electric power into light, is a good trade-off. Reasons might be high labor cost or nuisance of replacement (ceiling lights in an auditorium), or the involvement of personal or public safety (warning or exit lights). Closer to home, I would gladly waste a little power in the tail light of my car if it would never have to be replaced.

In ordinary use the main cause of demise of a bulb is evaporation of the filament, and that is a very steep function of the temperature. Bulbs manufactured to run a little cooler than normal are available (e.g., from Sears); typically about a 50% increase in life is claimed. The saver devices on the market go for much greater —in one case spectacular — extensions of life.

What we will need to know about tungsten filaments is readily found in handbooks, conveniently given in terms of voltage above or below normal, rather than temperature. The solid parts of curves A and B in Fig. 1 are reproduced from one such source.[1] Curve C, the life factor, is plotted from an empirical formula from the same handbook. For voltages near normal, the rule of thumb is that each 5% change doubles or halves the life.[2]

The saver examined by Bartlett and Geissinger was a wafer to be placed in the socket under the bulb base. It turned out to contain a *thermistor*, introduced in series in the circuit. The reason for using a thermistor instead of a resistor was puzzling at first, but the cleverness of it later appeared. The problem is to lower the voltage by about the same number of volts for a high or a low wattage bulb. A thermistor can approximate that. It is a semiconductor,[3] whose resistance goes *down* as its temperature rises. It heats internally when it carries a current, as anything having resistance does. Higher current (say when a higher wattage bulb is used) causes the equilibrium temperature to be higher, the resistance to be lower, and therefore (ideally) the voltage drop to be constant. In a test with bulbs of several wattages the voltage on the bulb was lowered by the amounts shown in Table I, column A. (It must be noted that the equilibrium temperature of the thermistor will vary with the type of socket, the ventilation, etc., so the data are only approximate.) For comparison, column B shows results that would be found by replacing the thermistor by a resistor. Clearly one resistor would not serve for all the bulbs, while the thermistor does so fairly well. The results for the thermistor indicate less life extension than the factor 4 claimed by the seller, but within the ball park.

I was able to examine three brands of quite a different kind of saver, a kind that employs a simple diode — a half-wave rectifier — to reduce the power delivered to the filament and therefore to reduce its temperature. Two were external wafers; in the third case the diode was concealed

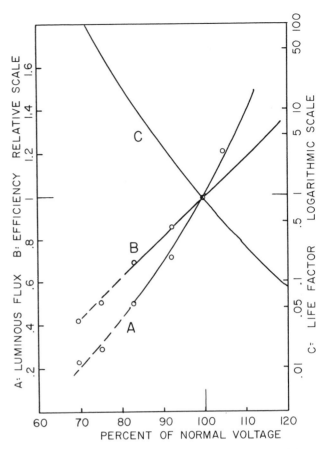

Fig. 1. Curves for light output, efficiency, and life, as obtained from handbooks and extended by experiment. All are on relative scales, "normal" values for a common bulb being 120 V and 700 to 1000 hours life. For curves A and B use the scales to the left. The scale for C is to the right.

inside the bulb base. We expect the effect to be the following: For any resistive load the power delivered is the time average of V^2/R. Half-wave therefore will heat the filament the same as would full-wave ac if the nominal (or peak) voltage were reduced by the factor $1/\sqrt{2}$ or 0.71. Curves A and B from the handbook do not go down that far, but they were easily extended. A light bulb, a variable voltage transformer, a photographic light meter and a meter stick did the trick. The reading on the light meter was made always the same, and the inverse square law was used to find the light output relative to that at normal voltage, which was taken as 120 V. The small circles and the dashed lines are the results.[4]

All the diode-type savers came out essentially the same. (The base of the third specimen was removed, to test it with and without the diode.) The light output was about 25%, as compared to operation of the same bulb without a saver, in good agreement with the curve at its 71% voltage point. The power consumed was measured to be 60%,[5] giving a relative efficiency of 40%. The claim of a factor of 100 in bulb life is not extravagant; it is consistent with curve C.

I was curious as to whether a half-wave load would confuse the watt-hour meter. To test that I turned off all

Table I		
Bulb	A. Voltage drop in thermistor	B. Voltage drop in 15 Ω resistor
15 W	8.0	2
25 W	7.7	3
60 W	6.3	7
150 W	4.5	16
300 W	4.0	29

other loads in the house, and ran two 250-W bulbs, first with the diodes in the same direction and second, with one reversed. The total load was the same; but in the first case half wave, and in the second case full wave. To my surprise, the meter charged the same to within about 2%, which was my accuracy. Smart meter!

Conclusions: The thermistor lowers the temperature enough to extend the life by a factor of 2 to 4, without much reduction in light output. That might be a good trade-off, for certain applications. There may be a dividend from the fact that the temperature comes up to equilibrium slowly (over a few seconds), which reduces expansion stresses and hot spots. For ordinary lighting there is a question as to whether the thermistor saves money. Besides lowering the efficiency, the wafer costs as much as about five 100-W replacement bulbs. The diode saver is a clear case of overkill. To obtain the same light as from a 100-W bulb in normal operation, a 400-W bulb would have to be substituted, which would consume 240 W.[5] Thus to avoid having to replace a 100-W bulb at the end of 1000 hours,

one would run up a bill in wasted power sufficient to buy a basket full of replacement bulbs. On the other hand, if you are using four times as much light as you need, you can install diode wafers and save both power and bulbs. Or you could just buy smaller bulbs!

References and notes

1. Keith Henney and Beverly Dudley, *Handbook of Photography* (McGraw-Hill, New York, 1939).
2. *Standard Handbook for Electrical Engineers*, 7th ed. (McGraw-Hill, New York, 1941).
3. A common type of thermistor is a sintered mixture of the oxides of Ni, Mn and Co.
4. Correction for the spectral shift with temperature was not possible. It is not believed to be serious. The light meter has broad response.
5. Not a simple 1:2 ratio, as might be expected for half-wave, because the resistance of the filament is less, at the lower temperature.

Metal Locators and Related Devices

A device that merely will give a beep when it is near a piece of metal is simple to make, but it will not satisfy today's well equipped treasure hunter. What is required now is discrimination between one kind of metal object and another — so the hunter can for example find the coins among bottle caps and pull tabs. The discrimination is tricky, and it is accomplished in several different ways. Here I will describe one of the ways — the one that seems to me to be the most sophisticated — and leave to a future column a simpler method, along with some words about the close parallels among methods of hunting treasure, detecting nuclear magnetic resonance, measuring the frequencies of resonant circuits and, according to recent ads, locating the studs in a wall.

All of the different metal locators sense the proximity of a hidden object by the object's reaction to, and effect upon, an alternating magnetic field that the locator maintains in the space near it. Because the frequency is not very high (the order of 10 kHz) the coupling between the searching device and the object is most like that between the primary and the secondary of a transformer, not like the reflection of radio waves as in radar. The common feature of metal locators is a coil in the form of a ring or pancake, 6 to 10 in. in diameter, mounted at the end of a wand so that it can be swept over the ground. Beyond that, systems differ somewhat in the way they sense and analyze, particularly in the way they achieve discrimination among objects of different composition.

For the particular detection system to be discussed here, the associated components are diagrammed in Fig. 1. A is an ac power source (a transistor oscillator) of fixed frequency, typically in the range ten to a few tens of kilohertz. It maintains a current in a driver, or transmitter coil B, located in the end of the wand near a receiving coil C. The alternating magnetic field it produces in the nearby space induces alternating current (eddy current) in the hidden object, D, which in turn induces ac in the receiving coil. The receiving coil is tuned (by the capacitor shown) to the frequency of the transmitter (and the eddy current) so it is very sensitive. In order that the transmitter not induce directly a strong current in C and mask the effect from D, B is physically oriented (as shown) so its coupling to C is weak. The dashed arrows a and b indicate the desired way the coupling goes: B to D and D back to C.

The possibility of discriminating between one kind of buried object and another lies in the fact that the *phase* as well as the amplitude of the eddy current induced in D depends on the nature of the object; mainly on its conductivity. (When we say phase we always mean phase with reference to that of the transmitter.) The eddy current in D, through the coupling of D to C, causes both a phase and an amplitude change in C; the combination being the "signature" of the type of buried object. A copper penny (good conductor) will give relatively a large phase shift and small amplitude change, and an iron bottle cap (poor conductor) will give the converse. The job of the rest of the circuitry is to interpret the signatures for the treasure hunter.

In Fig. 1, E is both a phase detector[1] and an amplitude detector. It compares the phases of B and C (arrows c and d) and sends to F a voltage (arrow e) that varies with the shift in phase of C as the wand passes over the hidden object. It also sends a voltage (arrow f) to F, which varies with the change in amplitude of C. F has a control by which the treasure hunter can play the voltages e and f against one another, either adding or subtracting, and can therefore null out the signal from a particular object, such as a bottle cap, if it is in a certain range of distance from the wand. Or the output from F (arrow g) can for instance be made to go down for the bottle cap but up for the coin. The readout can be by a meter, or a tone generator that varies in pitch or loudness. The latter can be biased so that, for example, the bottle cap is passed over in silence.

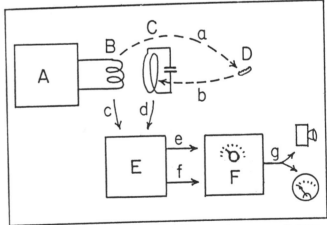

Fig. 1. Schematic diagram of a metal locator. A, ac source. B, transmitting coil. C, receiving coil. D, hidden object. E, phase detector and amplitude detector. F, mixer.

Only the main variables in one type of locator have been sketched here. There are many more possible adjustments, some that are useful and some that are frills. Examples of useful ones are adjustments to compensate the effects of currents induced in damp or mineralized ground, and adjustments to minimize the effect on the signature of the depth of the object below the ground or the end of the wand. The possibilities for knobs, bells and whistles are not limited.

Request for help

The different ways in which toasters time the on-period, switch the large (10 to 15 A) current, and eject the toast without throwing it are various and ingenious. There now is one that talks — calls you when ready. If we can collect enough schemes, new and old, we will run a column on them. Look inside yours and send the information.

We are hoping a camera buff will show up who can give us information as to how "auto-focusing" is accomplished.

Footnote

1. Circuits and techniques for phase detection or discrimination are available in electronics textbooks. To describe them here would take us too far afield.

A Spinning Top, Lenz's Law, and Electric Watches

A decade ago I bought a top that intrigued me because, although obviously powered in some way by a battery, it seemed free to wander at random in a shallow bowl. What further interested me was that there was no switch — no battery drain unless spinning. I cannot show a picture because I destroyed it finding out how it worked. But within it were some nice lessons about Lenz's law, and some illumination about a variety of electric clocks and watches that were popular at the time.

The spinner (Fig. 1a) was plastic, about 3/4 in. in diameter, and it contained a permanent magnet, aligned crosswise. It spun on a plastic bowl under which were a 9-V battery, a transistor and a coil that had two windings. A schematic is shown in Fig. 1b. As one end (say the S) of the magnet approaches (above) the coil, current is induced in winding A, in such direction as to make the base of the transistor (an NPN) go positive. That makes emitter-collector current flow, through winding B. The connections must be made such that the current in B goes around in the opposite direction to that in A (see arrows in Fig. 1b). The reason is seen from Lenz's law. It says that the current induced in A will be in such a direction as to repel the approaching magnet (just a statement of the law of conservation of energy: energy is dissipated in A, therefore work has to be done in pushing the magnet toward it). But we want the net force of the two coils to be toward them; therefore the current in B has to be opposite to, and greater than, that in A. We need not, in this connection, be concerned about the effect of A's magnetic field on B or vice versa, because those fields are very weak compared to that from the magnet.[1]

The beauty of the system is that energy is fed into the spinner no matter what. As the south pole approaches the coil it is pulled; as it leaves there is no force, because the current direction in A is reversed and the transistor is therefore nonconducting. When the north pole comes around, it approaches without force, but is pushed as it leaves. There is no preference as to clockwise or counterclockwise rotation. Neither can we confuse it by changing from an NPN to a PNP transistor: that would just call for a reversal of all polarities and directions, with no change in the end result. There is no drain on the battery when the spinner is not spinning, because the transistor (if silicon) does not start to conduct until coil A produces about half a volt. A germanium transistor begins conduction at about 0.1 V. (The device dissected had a silicon transistor.) A fast initial twirl of the spinner is necessary, to get over the threshold, but once spinning, the energy transfer rate is quite high — enough so that the spinner can spend most of its time wandering, only to pass near the center and get "revved up" once in a while. The spin rate is of the order of 1000 rpm. One further note: the coils, of very fine wire, and of about 2000 turns each, were wound over a small cylindrical iron core[2] (shown in Fig. 1a).

There was an interval between the eras of spring-wound and quartz watches when a method very similar to the one described was used to power the balance wheel.[3] The balance wheel carried a small permanent magnet and the coils (of incredibly fine wire) were mounted at the side. Of course the balance wheel oscillates rather than spins, but as pointed out, any motion of the magnet toward and

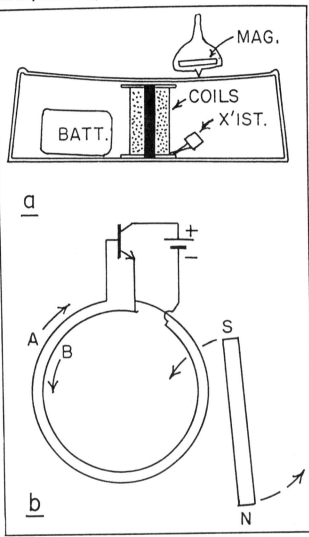

Fig. 1a. Cross section of the physical arrangement. b, The circuit. Only single turns of the coils A and B are shown. In the device each has the order of 1000 turns. The arrows indicate the directions of the currents and the rotation of the magnet.

away from the coil results in energy gain. The gear train of the watch was essentially unchanged, with the absence of a winding spring.

In the transition from the foregoing to the present-day quartz system, a curious improvisation was made by at least one of the major watch companies. It obviously required a minimum of change in tooling. A quartz crystal with an IC (integrated circuit) chip that produced pulses at the frequency of the balance wheel was simply connected to the coils near the balance wheel, and the transistor was eliminated. That gave the full precision expected of quartz, because the balance wheel was the captive. To stay locked in, the natural frequency of the balance wheel needed to be only very near to, not exactly equal to, that of the driving pulses. Now, for quartz analog watches, the balance wheel with its pleasant ticking is nearly a thing of the past — replaced by a simpler, ratchet type mechanism, that steps the gear train ahead typically once a second.

Editor's note: A spinning top that appears to work on the principle described here was advertised in *TPT* for Nov., 1982, p. 567. "...it spins for days on invisible wireless power. Secret scientific explanation included." We apologize if we have divulged secrets!

References
1. What induction there is from A to B is regenerative; it helps to make the current pulses in B stronger. The circuit may be recognized as the well known Hartley oscillator circuit. In this case it does not oscillate, partly because the Q factor is quite low, and partly because the transistor is not biased to favor oscillation.
2. For simplicity the driving process was explained without introducing the effect of a core. A core does not change qualitatively any of the results. It does enhance the drive.
3. Do-it-yourself clock movements of that type were offered in the familiar hobby catalogs at $6 to $9.

More questions

Can anyone give us the physics of how ink-jet printers work?

Robert Neff, of the "Good Reading" section of this journal asks how the airplane in-flight head-phones work — seemingly just a pair of plastic tubes. Where is the sound made?

Have You Picked a Lock Lately?

Back in the sixties when programmed instruction was to be the wave of the future, Basic Systems Inc.[1] put out as a promotion a free, 24-frame program "How to pick a lock." Not a trivial lock: a Yale-type 5-pin cylinder one. I flunked the test (on my own front door) but the good introduction to how a Yale lock works raised questions that caused me to look further, particularly to find how two keys—the private one and a master—can open the same lock. The local expert, Gerald Huller, Foreman of the Key Office of the University of Michigan, answered all of my questions. Since by design locks are not easy to probe into, a look at their inner workings may be appropriate for this column.

The cylinder lock in common use today is essentially that devised by Linus Yale in 1844.[2] But the basic principle he built on was far from new at the time. It is credited to Egyptian locksmiths at least 4000 years ago. Locks of that ancient type have had a continuous history: examples were found in excavations of the ruins of the city of Nineveh, and they are found to this day in parts of the Middle East.

Let's look first at the Egyptian lock. You will agree it is ingenious and simple. It is all made of wood. There is a sliding bolt, that is hollow and has several holes in its top (Fig. 1a). A piece called the staple, attached to the wall, contains 6 pins, which drop into the holes in the bolt, when it is in the matching position, to hold it

from sliding. The key is a paddle having 6 pegs, of such lengths and locations that when the key is inserted into the bolt, the movable pins can be raised to just the height that will let the bolt slide. The pins and pegs match, for a given lock, but their arrangement may be different for every other lock. With 6 pins and pegs, the chance that one key will work in another lock is sufficiently small. If there were faults with the Egyptian lock they were derived from the fact that the material was wood, and consequently bulky. The keys were as much as 3 ft long—not easily carried around.[3] The idea that has carried through from the Egyptian lock is that of raising a number of pins to just the positions that will allow one part of the lock to move with respect to the other.

Linus Yale changed the design to one of rotational instead of sliding motion, and he made the pegs internal parts of the lock instead of protuberances on the key. With the added advantage of metal as the material, the lock and key could be quite small. The lock, with the modern names for the parts, is shown in section in Fig. 1, b and c. The cylinder, containing the key slot, can rotate within the barrel. Holes, usually 5 in number, extend from the barrel down through the upper part of the cylinder and into the key slot. Each contains a driver and a pin, pushed from above by a spring. If the right key is inserted, the tops of all the pins are raised to the interface between cylinder and barrel (Fig. 1b), and the cylinder can turn (Fig. 1c). A wrong key raises the pin as in Fig. 1d. By a mechanical linkage the turning of the cylinder slides the bolt and lets the door open.

If you think about the above you will see how to make the lock open with two different keys, one a master and the other a private one. You just cut each pin into two segments (Fig. 1e) so there will be two heights at which it frees the cylinder to rotate. Being suspicious of such a simple solution, I got it verified by Gerald Huller. He commented further about a mistake that was made in earlier times, that of using the top of the top segment of each pin for the master key. That made it possible to file any private key down into a master, if the master could be borrowed long enough to copy its profile. Now one or more of the lower segments is used for the master.

You will not be forgiving if, after understanding how the lock works, we stop without telling how Basic Systems recommended jimmying it. You are told to maintain a gentle torque on the cylinder throughout the process. If, as is assumed, the pins fit a little loosely in the holes, the torque will turn the cylinder enough so that the holes in the barrel and cylinder will be slightly offset. One of the pins is now carefully raised by a hairpin in the key slot. If lucky, when the top of it reaches the interface, it will catch on the offset and be stopped (Fig. 1f). The torque that is being maintained is supposed to make some friction that will help to hold the first pin in position while you work on the next one. Good luck! Getting four pins to stay put while the fifth is teased may call for prayer or profanity. Basic Systems does not recommend which. Professional locksmiths start with the same technique, applying a torque to the cylinder to make an offset on which they can catch the pins. But there the similarity ends. They use probes that they fashion out of thin steel, and techniques of manipulation that get all of the pins caught at the same time.

An interesting question is why so many as five pins are used. If we assume, conservatively, that pins of only 5 different lengths are used, there will be over 3000 combinations. If a burglar were to try to enter by trying keys at random, he would have to have a 60-lb satchel of them. And that assumes all the keys would have the right groove pattern, which they would not have. Clearly, we must look to another reason as to why locks have so many pins. A guess is that it takes at least 5 pins to make it sufficiently difficult to pick the lock in a systematic way, along lines described above.

In the future there may be fewer holes in pockets due to carrying brass keys. A keyless system is making its debut in hotels: When a guest registers, he gives an

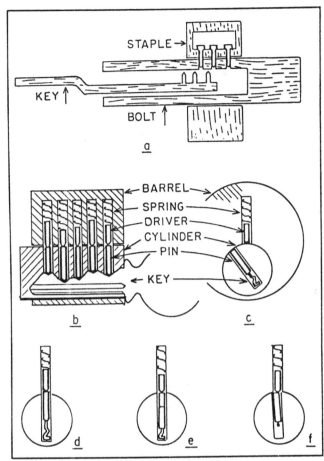

Fig. 1. a, Egyptian lock. b, section through a modern lock, showing the pins raised to the right level by the right key. c, end view, showing the cylinder turning. d, pin raised too far by the wrong key. e, pin cut into two segments to accommodate both a master and a private key. f, the picking procedure: hairpin raises the pin until it catches on the offset.

easily remembered number—birth date, street number, or whatever. It is entered into a computer. Outside the guest's room is a 0 to 9 touch panel that responds to the magic number and releases the door. When that guest checks out, the number is wiped out, so there is no chance for a past occupant to get back in, honestly or otherwise, as of course *can* happen, with the present key system. Another device, already familiar, responds to a plastic card containing coded magnetic material.

Notes and References
1. 2900 Broadway, New York.
2. First patent issued, 1844. Linus' son (Linus, Jr.) made improvements, patented in 1861 and 1865.
3. Walter Buehr, *The Story of the Locks* (Scribner's, New York, 1953).

How to Pump Blood without a Heart

A few of you may have special reason to appreciate pumps that can substitute temporarily for the heart. Other readers as well (and who knows who will be next?) will find some interesting points about pumps that have been developed to move blood. The requirements are quite special. First, of course, they should be completely closed to the outside environment, to prevent the possibility of contamination. Their valves or other mechanism should not smash or otherwise damage blood cells. A further characteristic (on which there is not full agreement) is that they should pump in pulses, to simulate the action of the real heart. There is a small class of pump designs that meet these requirements. Some are familiar in every-day applications: there is the squeeze bulb, seen by anyone who has had a blood pressure measurement or used an atomizer, and the diaphragm pump, used in the automobile as a fuel pump and as the compressor in some small refrigerators (Fig. 1, a and b). Both of these require valves, which may be flap-valves as shown, or small balls in cages. Another kind employs peristaltic waves, and requires no valves (Fig. 1c). In this the tube is flexible; a constriction moves from left to right, forcing the contents of the tube ahead of it. The constriction relaxes at the right, but before doing so another constriction starts at the left, so the content of the tube is trapped and can go only one way. Common to these three pumps are the facts that they are closed systems (no possible leakage to the outside) and that they require no rotating parts. The latter may explain why they are the ones that have evolved in nature: the first two are very similar to the heart, and the third to the intestine.

Sarns, Inc., a subsidiary of the 3M Company, located in Ann Arbor, is the world's largest supplier of blood pumps, and heart-lung machines. David Gayman of that company has been most helpful in explaining and demonstrating the pumps they make. Theirs is the peristaltic type,[1] arranged in a little more than a half circle (Fig. 1d). Two rollers move around, squeezing the plastic tube. It has to be a little more than a half circle so that one roller will close the tube before the other has moved off. The radius from the main shaft to the centers of the rollers is adjustable, and it is set so that the plastic tube is not squeezed shut completely; that is to avoid smashing blood cells. In operation the pump can be set to deliver the blood in pulses, just as the beating heart does. That is done by making the roller assembly turn a few revolutions rapidly and then hesitate. That capability takes on special importance at the end of surgery when the heart is beginning to beat and to take over part of the load. The pulsing of the pump is then synchronized with the heart by means of electrical signals taken from the heart. But even during the time the heart is kept completely inactive (by temperature or chemical means) there still may be some reasons in favor of pulsing the pump. One reason is simple and based in physics: when a fluid is forced through an elastic pipe (the blood vessels are elastic) the pipe expands, so that the resistance to flow decreases. The resistance to flow (pressure drop per unit length) has a very strong (inverse fourth

power) dependence on the radius of the pipe. The effect of this may go beyond simply getting a greater blood flow for a given pressure. Since the circulatory system comprises vessels of various diameters and degrees of elasticity, the distribution of blood among the different circuits will depend on pressure, and it may be more normal if higher pressure is applied in pulses (as from the real heart) than if lower pressure is applied continuously. Another reason for pulsing, but a somewhat conjectural one, David Gayman tells us, is that the body may possess sensors which, if the blood is not pulsing, will give warnings that the heart has stopped, and trigger countermeasures. We have not mentioned the chamber the blood goes through to be oxygenated, and we will not describe it here, except to say that if it is placed in the circuit ahead of the pump, the pulses will not smooth out and be lost in it.

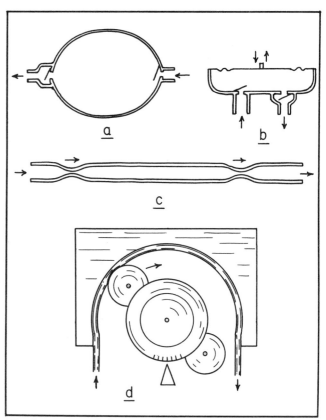

Fig. 1. a, squeeze-bulb pump. b, diaphragm pump. c, illustration of the principle of pumping by peristaltic action. d, peristaltic pump in rotary configuration, as used for pumping blood. The two rollers constrict the plastic tube. The dial in the center adjusts the distance between the rollers, to give the desired constriction or to accommodate plastic tubes of different diameters.

For demonstration, Sarns Inc. has a complete operating system that includes a life-size transparent plastic patient, with all of the internal cardiovascular plumbing. The heart-lung machine attached to this had, to my surprise, a battery of four peristaltic pumps, separately controllable. David Gayman explained that only one replaces the heart action; the others are used to infuse solutions (for example a drug/cold solution that stops the heart) or are used for suction, venting, or other blood-reclaiming functions. Of course to go with all this there are digital and cathode-ray readouts for monitoring everything. It is an experience to see it all in operation.

Some numerical data may be of interest. The outer diameter of the roller assembly is 6 in., it is turned by a 1/8-hp motor, and its speed of rotation is controllable from 5 to 250 rpm. The tube is polyvinyl chloride. Tubes of outside diameter 1/4 to 3/4 in. are accommodated by an adjustment of the roller radius. The fluid volume external to the patient is 400 to 600 ml, so that requires the addition of fluid when the system is connected. Saline solution, not blood, is added. That much dilution of the patient's blood is tolerable.

Reference

1. There are blood pumps on the market that operate on other principles; in particular, one that produces a vortex and pumps by centrifugal force. It is made by Biomedics, Inc.

Light, Dark, or Burned?
Two Automatic Toasters

Under the sleek cover of a toaster (assuming you can figure out how to get it off) you will find the insides are more complex than the simple idea of toasting a piece of bread would suggest. You may be impressed further that such an intricate "mousetrap" can last the better part of a lifetime. The several basic actions required in the cycle are accomplished in different ways in toasters of different brands and ages. I did "reverse engineering" on only two kinds. I invite information on other kinds, and offer to pass it along in a future column. Among the functions the toasters dissected were able to perform are: a large current (12 to 15 A) is switched 10^5 or more times with little erosion of the contacts. The on-time is determined mainly by the surface temperature (and/or darkness) of the toast. At the end the toast is raised in the slots, not tossed into the air. The darkness is controllable by a knob. Only in one was there a "panic button" — a way of stopping the process in case of smoke, other than by pulling the power cord. Here is what I found. The sketches will indicate just the key elements, schematically, to show the principles in a simple way. Pivots will be indicated by small circles. Springs that return parts to their normal positions will have to be imagined. I cannot avoid having the description read like that of a Rube Goldberg machine[1]; it's in the nature of toasters.

The first is an old (maybe 20 years) Sunbeam. The main moving mechanism consists of a pair of horizontal bars, one on each side, inside the case; the near one shown as a in Fig. 1. Attached between them are two cradles, b and b', that support the bread. The bars, held by stiff wires c and c', can swing down to the position shown by dashed lines, lowering the bread in the slots. To begin with, the bar structure is held in its up position by a linkage from d to the center (e) of a short piece of nichrome ribbon, shown at the bottom of Fig. 1. When bread is dropped into the slots, one of the pieces depresses a finger, f, which starts the heater current. The nichrome ribbon is in series with the heaters, so it gets hot and expands to the position shown dashed, allowing the bar structure to go down and lower the bread. (Rube, we need you!) When the toasting is finished and the current stops, the nichrome strip shrinks, lifting the bars and the toast. What remains to be told is how the finger f starts the current and how the cycle is stopped when the toast is just right.

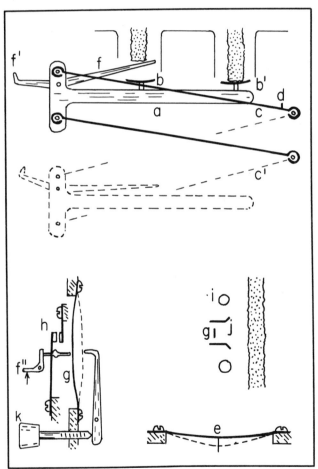

Fig. 1. Schematic representation of the essential parts of an old Sunbeam toaster.

When the finger is depressed, its other end f' goes up, pushing on f'' (lower left in Fig. 1), which pushes a pin that makes a bimetal strip, g, snap[2] to the position shown dashed, allowing the main contacts h to close and start the heating. The bimetal is located so that it "looks at," and receives heat from, the toast. When it gets hot enough it snaps back to the original position, opening the contacts quickly, so there is little arcing. To make the time at which the bimetal snaps back depend as much as possible on the surface temperature and darkness[3] of the toast, the bimetal is shielded from the direct radiation from the heaters (which are vertical nichrome coils, i) by blinders j. The adjustment for darkness of the toast (knob k) determines the position (dashed) of the bimetal, and that changes its threshold for snapping back.

The second toaster is a Proctor-Silex, a model currently being sold. It is not less complicated than the first; just different. The feature common to the two is a bimetal element that senses the condition of the toast. The bread is lowered in the slots by pushing down (against a spring) a handle (Fig. 2, ℓ) at one end of the toaster. Three things happen: the pivot of hook m comes down so that it latches to n and holds the handle (and bread) down; o is pushed down, closing the two contacts p, p', starting the current in the heaters; the piston in cylinder q is pushed down. When the toast is (supposedly) a golden brown, a bimetal strip r (located close to one of the pieces of toast) curls, pressing a sensitive switch s, causing the electromagnet t to unlatch u, tripping the series of latches and letting the handle and the toast go up. The (leaky) piston-in-cylinder slows the rise just enough so the toast is not thrown out, but the rise is still fast enough to let the contacts open quickly, without arcing. The bimetal strip is not of the snapping type; that is not necessary, because before the sensitive switch opens, the main contacts have already opened, stopping all current. To determine how brown the toast shall be, the sensitive switch is moved closer to or further from the bimetal by a lever, v. Provision is made for ending the toasting at any time, in case of smoke: the hooks are shaped so that the linkage holds if undisturbed, but so that it will be tripped by an upward force on the handle. The circuit of the heaters in the two slots, the electromagnet, the double main switch, the sensitive switch and the bimetal is shown at lower left, Fig. 2.

Toasters have remained essentially mechanical gadgets over a long period. Only the clock-like escapement for timing seems to have given way to the bimetal strip. We wonder if the time is ripe for the integrated circuit chip to take over.

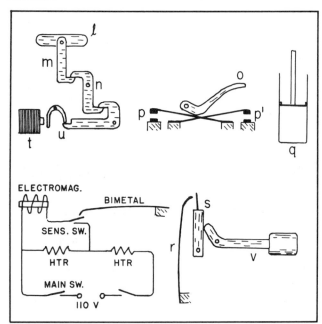

Fig. 2. The essential parts and the circuit diagram of a current model toaster (Proctor-Silex).

Footnotes

1. For an example of the Rube Goldberg principle, see "From a student's laboratory notebook," Phys. Teach. **17**, 377 (1979).
2. The bimetal strip, widely used, consists of two sheets of metal of different temperature coefficients of expansion, welded together to make a sheet of total thickness typically 0.010 to 0.020". It becomes curved when heated. In many applications snap-action is required, usually to open and close electrical contacts quickly. Snap-action is achieved by making the sheet concave — dished — or by fastening a strip of the bimetal at both ends. There is a familiar concave circular disk of bimetal which, if dropped onto a hot surface will reverse curvature with a snap and jump high into the air.
3. As the surface darkens, the absorption and emission of radiant energy increases.

Your (Not Always) Friendly Traffic Signals

We deal with traffic signals every day of our lives, yet few of us know much about the workings of the system. In every town of any size there is a "nerve center," most likely in an out-of-the-way building, not in City Hall. There, chances are you will find a friendly fellow with physics and electronics training, who will be pleased to explain things. The fellow I found in the Ann Arbor center was Gary Fitzgerald. We started with the mainframe computer. It receives, continuously, traffic flow information from sensors under the pavement at numerous points in the roads between intersections. One of its outputs is to a 10 x 10-ft wall map of the city streets, on which there are small indicator lights at all of the 130 intersections that have traffic signals. The indicators come on and off as the actual signals change; a fascinating sight to watch. There are other lights to indicate trouble. When I watched it, four or five of those were blinking. At the punch of a button, printouts can be had for the traffic, by the minute, hour or day, for any of the sensor points. The main function of the computer is to process the information and send orders back to the traffic signals, to modify their time intervals of red, green, etc., according to the requirements of the hour. Information from the sensors and to and from the signals is via lines leased from the phone company.

At each intersection the repeating sequence — green, yellow, red, left turn, etc., is executed by circuitry in a box (typically 1 x 2 x 3 up to 6 ft in size) mounted near the ground on one of the poles that hold the cables on which the array of lights is suspended. If you listen you may hear the clunk of contactors in the box every time the lights change. But you may find that some boxes are silent. That is because we are in a transition period in which solid-state controllers are replacing the electromechanical ones. In Ann Arbor the change is about a quarter complete. The change is slow because the solid-state controllers run from about $3000 to $6000 apiece. But they are worth it because they have "brains," by which they improve traffic movement. In fact we may soon have come full circle: the solid-state controller's management of the traffic now comes almost up to that of the cop in the middle of the intersection, of olden times. (At higher investment but lower maintenance!)

The old and the new controllers differ so much in what they can do that they will have to be described separately. I will take the old, and (in my town) more numerous ones first. At the heart of the electromechanical controller is a motor-driven multicontact rotary switch that determines the sequence of red, green, etc. The relative time intervals can be set manually by movable tabs. The rotary switch controls heavy-duty contactors which turn the lights on. The timing set by the tabs can be modified in an important way, by wire from the central station: the motor can be ordered to pause, for a specified length of time, each time it turns on the green light, for any of the directions of traffic. By that means the green interval for the direction of heaviest traffic can be stretched to suit the changing demands of the traffic flow. (To avoid executing changes on short-time fluctuations of traffic, the computer averages the flow for about 15 minutes.) A further use made of the pause is to set the phases of the signals along a long thoroughfare, so that, say, cars going 30 mph in one direction in the morning can stay in phase with the green, and conversely in the afternoon. Needless to say, the motors that turn the rotary switches are the type that run in synchronism with a 60-Hz power, so once shifted to the right phase they remain there. Phasing is generally not ordered on the basis of current counts from the sensors, but on longer time averages, because rush-hour traffic repeats day to day. The same is true of the flow at some individual intersections.

The solid-state controller, having a brain of its own, needs, and gets, little instruction from the central station. Where one is installed, sensors under the pavement in all lanes feed their counts to it. The controller calculates and allots the relative green intervals in the optimum way. Further, it can avoid holding a direction open after the pulse of cars has ended, and it can skip the left turn interval if the sensor says there is no car waiting for it. Only when cooperative behavior among controllers is required, namely for phasing along a boulevard, are instructions via the telephone lines needed.

The under-pavement sensors for counting cars are interesting, in view of the fact that metal locators were described in a recent column.[2] Sensing a ton or more of iron should of course be a cinch compared to finding a lost penny in the sand. Earlier, taking advantage of the iron, magnetic pickups were used. The trend now is to a system very much like that of the penny locator. A wire coil under the pavement, tuned by a capacitor to a frequency of around 100 kHz is continuously energized by a crystal-controlled power source. The "loading" and consequent reduction in amplitude in the coil-capacitor circuit is what gives the signal that a car is above it. The computer or controller takes care of making one car count only once. The wire coil consists of only two turns, is 4 ft wide by from 6 to 70 ft long, and is about 2 in. below the surface. The two turns are tucked into a saw cut in the pavement, later sealed with hot asphalt.

Recalling another of my recent columns,[3] I asked about the life of the filaments in the lights. The bulbs run at 115 V, 60 to 135 W, and look like the ordinary spherical ones found in the home. But they are made special, in that the filaments run at a reduced temperature. (Curves given in the earlier column showed that great extension of life results from lowering the temperature.) The bulbs are routinely replaced once a year, which is well within their lifetimes; therefore signals seldom go dead because of bulb burnout.

My questions brought out some other interesting facts. The red, yellow, or green disk you see is a variation on the Fresnel lens. It is a disc of colored glass. (All bulbs are white.) On the rear surface is a matrix of small lenses, that make a beam of the light. On the front surface is a series of vertical grooves (prism-like) that spread the beam horizontally just the right amount. Light is cleverly conserved. The glass in the left turn signal is different: it does not spread the beam horizontally. It is designed so that the driver in the neighboring through-lane will not see it and mistakenly turn. To service the 130-intersection signal system there are five persons in the field crew, using two trucks equipped with "cherry pickers." The "command center" has five employees. They work only in the day, leaving night operation to the computer.

The ultimate test of the variable-interval system, Gary Fitzgerald said with evident satisfaction, comes when the Michigan football stadium lets out 105 000 homeward-bound fans. Based on data gathered from many past game-days, a computer program has been developed that does the whole job with maximum efficiency — while Gary watches it happen on the map with the blinking lights.

References

1. Topic suggested by Gerry Rees, of Washtenaw Community College.
2. Phys. Teach. 22, 38 (1984).
3. Phys. Teach. 21, 606 (1983).

Physics in the Copy Machine

Since 1939 when Chester Carlson applied for his first patent on a simple system of copying onto plain paper by electrostatics, machines have become very complicated, with many variations and improvements on the original method. Each variant is, of course, promoted by a company as being the best. We don't think you will want to read a full treatment; we think you will be more interested just in some of the clever uses of simple physics principles. Of particular interest (and at the heart of the copying process), are the ways in which electric charges, or particles carrying charges, are caused to transfer from one surface or medium to another at each stage. Let's go briefly through the main steps of the copying cycle, sticking to the physics and not trying to go into the particular ways the different machines do the steps. Most of the information to follow was supplied by Professor D. J. Montgomery of the College of Engineering of Michigan State University, and Dr. R. W. Gundlach of the Xerox Corporation.

The copying system begins with a *photoconductor*. A few substances, notably selenium, arsenic, and tellurium (and their alloys) are photoconductors – insulators in the dark, but conductors of electricity (although poor conductors) in the light. In the copying process as worked out by Chester Carlson and as used, in improved forms, to this day, a thin layer of photoconductor[1] on a metal backing is first given a uniform surface electric charge. This is done in darkness, so the photoconductor is a good insulator, and the charge is retained. Then the image of the page to be reproduced is projected, optically, onto the charged surface. That allows the surface charge in just the lighted areas to leak away to the metal backing and to ground, so a "latent image" is left in the form of a distribution of surface charge. The next stage is to "develop" the latent image; that is, to render it a physical, rather than an electrical, image.

Carlson developed the latent image very simply, by wafting an airborne dust of black particles (now called toner) over the charged surface. The particles were attracted to the charged areas. That method is still used for some special purposes, but more sophisticated methods of delivering the toner have been developed. One of these is called a "cascade" method. It employs intermediary particles or "beads" to carry the toner.[2] The beads – in some cases glass – of up to about 500 μm diameter, have toner particles clinging to their surfaces. The toner particles are much smaller: spheres or spheroids of about 10 μm in size. They are plastic, pigmented with carbon black. The beads roll, or cascade, over the photoconductor surface, and *in the charged areas* toner particles are attracted off the beads and retained by the photoconductor (Fig. 1a). In a related system, the carrier beads have magnetic cores and nonconducting surfaces. As in the other system, they carry toner particles on their surfaces. The beads with their coats of toner are picked up by a magnet, in such a way that they form chains, or whiskers, making in effect a soft brush (Fig. 1b). The brush wipes over the photoconductor surface, and as before, toner particles transfer to the photoconductor in just the charged areas.

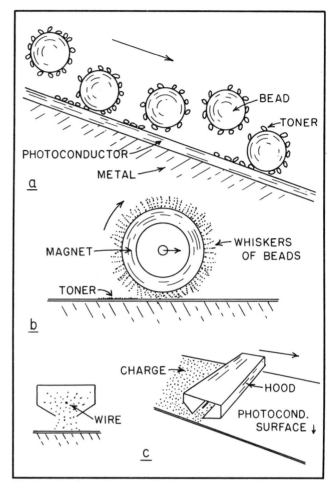

Fig. 1. a. Beads moving over the photoconductor surface, carrying, and transferring, toner particles where the surface is charged. **b.** Cylindrical magnet (as long as the width of the photoconductor) bewhiskered with beads carrying toner, rotating and brushing the surface of the photoconductor, leaving toner where the latter is charged. The motion indicated is relative: the cylinder can traverse the photoconductor, or the latter can do the moving. **c.** Corona device for ionizing the air. The wire is about 0.1 mm in diameter, and at a potential of 5 to 10 kV positive with respect to the grounded hood and the photoconductor. The indicated motion is relative: the wire and hood, or the photoconductor, may move to cover the area.

The processes just described make a physical image in toner particles, on the photoconductor plate. Following that, the toner image has to be transferred to a piece of paper, and the particles have to be fixed to the paper so they will not rub off. The transfer is made by rolling the paper against the photoconductor. The fixing is done (commonly) by running the paper through hot rollers, so the toner particles melt and stick to the paper. (Needless to say, when that method is to be used, the toner particles have to be made of a thermoplastic of suitable melting point.) At the completion of the whole cycle, the photoconductor surface has to be brushed free of left-over toner particles, to be ready to receive a new surface charge. Such is a cycle from the viewpoint of a physicist. A mechanical engineer would see it differently: complex machinery for feeding paper, moving many parts, even giving change for a quarter. A company person would see that important variations on the means of executing each step have been left out.[3]

Through the foregoing description you should have been wondering how the toner particles are to know which surfaces to leave and which to attach to, at each of the stages. That problem has called for clever planning, involving the most interesting physics of the whole process. The movements of the particles are dictated, of course, by electrostatic attractions. A particularly interesting (to a physicist) bit of very old electrostatics lore that enters is the *triboelectric series*. That, in analogy to the better known electromotive series, is an arrangement of the various nonconducting substances in the order in which each will take electric charge (say positive) from another, when rubbed together.[4] It is involved in all cases of "friction electricity." The common example is that cat's fur takes positive charge from glass, leaving the glass negatively charged. We shall see how the phenomenon is used in the cascade process.

The charge given to the photoconductor surface is to be positive, for reasons soon to appear. Charges of both signs, as ions of nitrogen or other molecules, are produced by a corona discharge (Fig. 1c). The wire in the corona device is positive, so the positive ions move away from it and to the photoconductor, which is at ground potential. If the toner particles are to be attracted to the areas of the photoconductor that remain positive, they must be negatively charged. That is where the triboelectric series comes in. The carrier beads are made of, or covered with, a substance (higher in the triboelectric series) that will take positive charge away from the toner particles. When the beads and toner are stirred together, the toner particles become negative (leaving the beads positive), so the toner clings to the beads. But the unexposed (charged) areas of the photoconductor are positive so they steal some of the toner from the beads. At the final stage, to help the paper take the toner away from the plate, the roller behind the paper is (in some systems) given a positive potential. There is an element of luck in the foregoing sequence, in that the final copy ends up a photographic positive with respect to the original — dark for dark.

Footnotes

1. They are to be distinguished from *photovoltaic* substances, which produce voltage and therefore can deliver electric power when exposed to light, as for example the silicon panels that capture power from the sun. Still another device is the *photoelectric* cell, in which electrons are ejected from a surface by light, typically in a vacuum.
   Carlson used selenium as the photoconductor. Although now there are alternatives — including some organic compounds — selenium and its alloys have an advantage in toughness against wear and bending. Typical thickness of the coating, on the metal backing, is about 50 μm.
2. Airborne toner is attracted according to the electric *field* near the photoconductor. That tends to give deposition only at the edges of areas of uniform charge. The cascade and magnetic brush methods of delivery follow more nearly the *charge* distribution.
3. Here are some variations. Photoconductor may be on a plate, cylinder, or endless metal belt. Fixing may be by softening the toner with a volatile solvent, or simply by pressure. Toner may be delivered by suspension in a nonconducting liquid. The photoconductor may be disposable — in a special paper. A fiber-optics system of transmitting an image of the original to the photoconductor has been developed. A recent review of methods used by various companies is in *High Technology*, May 1983, p. 49.
4. The triboelectric series predates the electromotive series by a long time; can be traced to Rostock University, around 1750, Professor Montgomery tells us.

Ionization Smoke Alarms

The little smoke alarm the diameter of a tea saucer, that wakes you up when your house is on fire, is called an ionization detector. But the role played by ions is quite different from what a physicist would first think of. I learned that from the source of most of the information that will follow: Chris Zafiratos, of the University of Colorado. His words describe his (and my) surprise: "When I first heard of ionization detectors for fires I was aware that fires *produced* ions. I inferred that these devices sampled air by means of a small blower and passed it between electrodes that were at substantial potential difference. The electric field between the electrodes would draw the ions out, and they would constitute a small current which when amplified would sound the alarm. I was, however, concerned by two things. First, how would ions survive a long trip from a distant fire? Second, why were radioactive materials involved in these detectors? To answer these questions I volunteered to seek further information."[1]

Chris goes on as to what he found: "A radioactive alpha particle emitter, Americium-241, produces ions in a small chamber that has louvres through which air passes, and two electrodes at a potential difference (Fig. 1a). A steady current flows due to the ionization of the air by the alpha particles. When combustion products enter the chamber, the ions attach to the particles (smoke or invisible ones) and are nearly *immobilized*. Thus the current through the chamber is sharply reduced, at least for many seconds. With appropriate circuitry, that sets off the alarm."

It must be remarked here that the system of making ions locally and letting them be immobilized by the particles from a distant fire gives a far greater change in current than what would be obtained by simply collecting ions from the fire. But even so, the current change is extremely small; it is in the range of 10^{-10} A.[2] How to sense that, reliably, is the second problem that has to be solved. The device that is used to make the first step of amplification must require almost no current at its input, that is to say, it must have a very high input resistance. After that first step is made, the rest of the circuitry leading to the ringing of bells, sending messages to the fire station or whatever, is standard electronic practice. For that part we can safely "leave the solution to the student."

Two devices are used for making the initial step. The first, very ingenious, dates back nearly half a century to a Swiss physicist Ernst Meili, who was engaged in using an ionization chamber for the detection of particulate matter in the air in mines.[3] He adapted a cold cathode (gas discharge) tube to the problem. It is used to this day, although not in the popular detectors for the home, for the reason that the cold cathode tube is not suited to battery operation. It requires a high voltage (~ 150 V).

The cold cathode tube is very similar in operation to the tube used in elevator touch panels, which we described in these columns.[4] The main electrodes (Fig. 1b) are kept at a potential difference below that which will start a discharge. A third electrode, a point close to the cathode, can, when given a positive potential, increase the field locally near the cathode sufficiently to start the main discharge. Once started, it keeps going, with current high enough to operate a relay that will ring the firebell. Simple. Until the moment of discharge, no current flows from the point; the input resistance may easily be 10^{13} Ω, and that may in fact be the limit set by leakage over the external surfaces, rather than by current from the point.

For inexpensive smoke alarms for the home, the necessity for high voltage (needed for a gas tube) is avoided by the use of one of the marvels of the transistor age, the IGFET (insulated gate field effect transistor) also called MOSFET, the MOS standing for metal oxide semiconductor (Fig. 1c). As seen in the figure, the conventional representation shows the control gate as isolated, so it controls by its electric field, not by current flow into the transistor. Its input resistance can be as high as 10^{15} Ω. Needless to say, surface leakage may be predominant, and make the effective input resistance lower. The MOSFET works on a 9-V "transistor radio" battery, and that allows the smoke alarm to be self-contained. I found the drain to be 18 μA during quiescence and 60 mA when the alarm was sounding. Battery operation is not without hazards: Murphy's law says the battery will be dead just when the fire occurs. To counter this, the circuit is designed so that when the battery is low, the noise-maker (*horn*, it is called) will give short blasts at long intervals. The one I tested started the warnings when the voltage was lowered to 7V (from normal 9 V). I made one other test: I found that the voltage change at the *output* side of the MOSFET, when I blew pipe smoke into the chamber, was several volts — a healthy signal for further amplification leading to the horn.

The system described has other applications. Recently a furnace service man came to our house, and as a check on possible leakage in the heat exchanger, he held a small device in front of a hot air outlet. It chirped at intervals of about a second. An increase in the chirp rate would have indicated combustion products. I wrote to the manufacturer for technical information, but received no answer. Trade secrets, no doubt. A good guess is that the device is the same as the smoke detector we have described, except for the chirp.

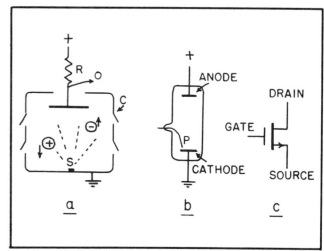

Fig. 1. a., the ionization chamber. C is the louvred enclosure, R is an extremely high resistance ($10^{12} - 10^{13}$ Ω), and S is the alpha-particle source. Ionization current maintains a voltage drop in R. When combustion products enter the chamber, the ion current decreases and 0 changes potential toward positive.

b, the cold cathode tube, showing the point, P, near the cathode. Positive potential on the point initiates the anode-to-cathode discharge, whose current is large enough to operate a relay.

c, conventional representation of a MOSFET. The gate is shown as isolated from the other elements, to indicate that it controls the source-drain current by its field; not by current flow. When connected to 0 of the chamber, the MOSFET must be biased so that the source-drain current is normally cut off. There are several possibilities: the MOSFET may be either p or n type, or the polarity of the ionization chamber may be inverted from that shown.

References

1. Consultation with Azmi Imad, Director, Environmental Health and Safety Dept., University of Colorado. Reprint of a talk given by J. E. Johnson, Pres., Pyrotronics, Inc., and an applications manual. Descriptive material from Pyr-A-Larm Co.
2. The current change is estimated as follows: The radioactive sample is the order of 10 μC — weak enough that a license is not required. 10 μC gives (3.7×10^5 disintegrations)/s. The alphas have a range of about 4 cm, well suited to produce ionization in a small chamber. Their energy is about 5 MeV, and in air they produce about 1 ion pair per 30 eV. But not all of them go out into air of the chamber; many go backward into the substrate, or lose part of their energy in the substrate. A reasonable estimate might be that a third of all the energy of the alphas gets used in making ion pairs in the air. When collected, those ions amount to about 3×10^{-9} A. If a change of, say, 10% or less is to activate the alarm, we are in the range of a change of 10^{-10} A.
3. J. E. Johnson's talk, Ref. 1.
4. Phys. Teach. 21, 402 (1983).

Automatically Focusing Cameras

Step by step the amateur photographer has been relieved of the necessity (and in the view of old-timers, the challenge) of having to estimate and set the exposure time and lens aperture. Those adjustments are now fully automatic for both natural light and flash, in many cameras. In the last few years the push has been toward relieving the operator of yet one more function: focusing. What can come next? The only variable left is the choice of the subject at which to point the camera. Although more poor judgment is to be found in that choice than in all the others combined, automated help there is not easy to imagine. So the end of things to automate may be at hand.

The advent of autofocus has brought an interesting change in the mix of camera types advertised. In a period before that, the single-lens reflex type had pretty much displaced the range-finder type in the expensive camera category, albeit at the disadvantage of extra bulk and weight. Now that the focusing problem can be made to disappear, there is a resurgence of interest in the "compact" type, which in fact goes back to the shape and size of the earlier range-finder cameras.

As we saw in the early development of phonograph records, televisions, sound and video tape recorders, and other devices, in each case a number of schemes had to run their courses before one prevailed as the best. Judging from current advertising, autofocus systems seem to have about completed that shakedown, with convergence toward one system: so-called "active infrared" ranging. It is "active" because infrared radiation (IR) is sent out to the subject, and received back, by the camera; it does not depend on existing light. That is an important factor, because it ranges as well in the dark, for flash pictures, as in daylight. (A substantial fraction of pictures are taken by flash. Even in daylight!)

It is not possible here to give details of how auto-focusing is accomplished in various makes of camera. I can at best describe the physics principles, and the main variations in the way they are applied. Finding out that much was not easy, since most of the IR autofocus cameras are imported. And reverse engineering (dissection) was not practical on $175 cameras! My best source was Dr. Forrest Strome, of Eastman Kodak Co. Some articles and letters were helpful,[1] as were salespersons in stores.

The IR autofocus systems build on the well known range-finder principle, which is, in surveyor's terms, triangulation (Fig. 1a). In the typical system the lens is moved manually, and that, through a mechanical link, rotates the mirror shown at the right in the diagram. When the two images are seen to coincide, the camera is in focus. The focusing is specific: that is, a small area in the center of the field of view is brought into focus.

In the autofocus system, a narrow beam of infrared is projected out of one of the windows of the camera, to illuminate a spot in the center of the field of view. The IR-illuminated spot is "seen" by a sensor[2] that looks out through the other window. In several different ways that I will describe, the angle between the outgoing and ingoing paths is made to determine the position to which the lens

shall be moved, to be in focus. The lens is moved by a small motor.[3] When the button is pressed, the motor starts moving the lens outward from the camera (from the infinity position). When the in-focus position is reached, the motor stops and the shutter makes the exposure. Of course all of these events are directed by a brain: an IC (integrated circuit chip).

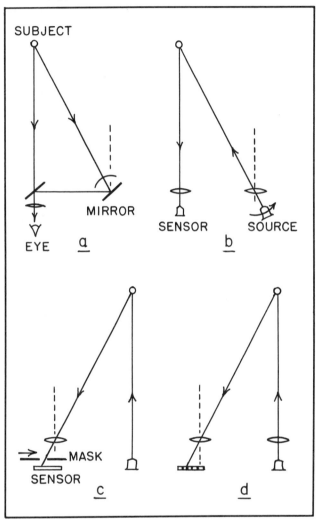

Fig. 1. Four range-finder configurations. The dashed line in each case indicates ranging on a subject at infinity. The lenses shown are the range-finder lenses, not to be confused with the picture-taking lens which is not shown.

a: A conventional visual, coupled range-finder. The mirror on the left is half-reflecting, and is fixed. The mirror on the right is full reflecting; its rotation and the movement of the picture-taking lens in and out are linked mechanically. When the eye sees the two images in coincidence, the camera is in focus.

b: Configuration for infrared ranging, in which the sensor remains fixed, while the beam from the source is swept in direction, by means of a mechanical link to the main lens.

c: Infrared ranging using a fixed source, a large-area sensor, and a mask that moves, by linkage to the main lens. What moves is variously described as an edge, a mask with a slit, or a narrow bar.

d: Infrared ranging using a fixed source and a segmented sensor whose segments are connected individually to the integrated circuit chip.

The variations in the systems are in the details of how the angle between the outgoing and ingoing paths is translated into a setting for the lens. The principal ways are shown by the sketches in Fig. 1, b to d.[4] The first, Fig. 1b, is the most like the visual coupled range-finder system. The sensor "sees" only a small area straight ahead (center of the field of view). When the main lens is moved outward by the motor, the IR source is moved (through a linkage) so that its beam sweeps from straight-ahead (dashed line in the figure), toward the left. When the beam strikes and illuminates the small area that the sensor is seeing, the motor is stopped and the exposure is made. In a similar system (Fig. 1c), the outgoing IR beam direction is fixed, and a mask with a slit or hole in it moves across an extended IR sensor. The slit position and the lens position are linked. When IR comes through the slit, the motor is stopped and the exposure is made. In a third system (Fig. 1d) the only moving part is the lens. The sensor is divided into segments, typically five. When the button is pressed, the motor moves the lens outward; the IC chip tells it when to stop, depending on which segment is receiving or has received IR, and the picture is taken.

From the physicist's point of view there are two interesting questions not yet answered: (1) How is the rather weak IR radiation that reaches the sensor distinguished from the ambient IR and other "noise"? (2) How precise is the focus, compared to that which would be achieved with a manual range-finder or a single-lens reflex system? I have only partial answers.

The first is accomplished by a sort of autocorrelation. The transmitted IR is modulated at a frequency (50 kHz in one of the cameras), and the circuit following the sensor is tailored to recognize only that frequency. A variation is to send out a single, short, intense pulse of IR, and to have the circuit gated to the same time interval. That system can work only with the system having the segmented sensor; the IC chip has to remember which segment received the IR, until the lens has had time to move to the corresponding position.

As to how precise the focus: it undoubtedly is better than that achieved by the casual amateur picture-shooter, but not as precise as what can be achieved by the veteran photographer, using manual focusing. Evidently to keep the demands on the autofocus system within its capabilities of precision, all the cameras (whose specifications I have seen) have lenses of short focal length (35-38 mm), f numbers no less than 2.5, and minimum focusing distance no less than 3 ft.

Some further remarks may be in order. As indicated earlier, the autofocus system works on a spot at the center of what is seen in the viewfinder. That makes a problem (easily solved) if the object you wish to be in sharp focus is not in the center. The procedure is first to aim the camera so the object to be sharply focused *is* in the center, and to push and hold the button halfway down. That fixes the focus and exposure, but does not trip the shutter. Then you can re-frame the view and push the button the rest of the way down. Finally, it should be mentioned that, as you probably have assumed, the usual automatic light metering and exposure setting, as well as automatic aperture setting for flash are included in the autofocus cameras.

In summary, whether or not you are an expert, autofocus cameras can do some things for you that you cannot do for yourself. And vice versa. They are not able, as presently designed to give you the sharp focus with low f-number and/or long focal length lenses, which you can get so well by eye. But when it comes to speed of focusing, they win hands down. They focus in a fraction of a second, and therefore can get many shots that would be impossible with manual operation.

References

1. "Six autofocusing compact 35s" *Popular Photography* Nov., 1983, p. 94. Letters and specification sheets from: Mark Carmany, Pentax Corp.; John J. Jonny, Minolta Corp.; Jeffery M. Karp, Ricoh Corp.; Scott J. Leibow, Nikon Inc.
2. The sensor is an infrared-sensitive silicon diode. The source is an infrared emitting diode (IRED). They operate at about 900-nm wavelength. Both devices can be purchased from Radio Shack, at small cost, for home experimentation. That is recommended.
3. Typically powered by two AA cells. The same cells can power automatic rewind, if that is included. In some instances, to reduce the power needed to focus, only one element of the lens is moved.
4. Diagrams of the principal configurations were supplied by Scott J. Leibow of Nikon Inc.

Halogen Lamps

If you have had to replace the bulb in your headlight or slide projector in recent years, you know from the high price that they are not made the way they used to be made. Many of the high performance (high light output from small size) lamps now work on the *halogen cycle*. These lamps contain a small amount of the vapor of a halogen: iodine or bromine. That works wonders for the operation, but to provide the conditions that are required for the chemical reactions, changes have to be made in the construction that run the cost up. That is why ordinary consumers like ourselves meet them only when we buy special-purpose lamps, not when we replace household bulbs. We will go into the halogen cycle later; first let's take note of what it does — why it's worth (in some cases) the extra price.

The halogen cycle goes after the main problem in the conventional tungsten filament lamp: the evaporation of tungsten from the filament to the inner surface of the glass envelope. The loss from the filament finally results in burnout, and the deposition on the envelope gradually reduces the intensity of light that gets out. The halogen cycle helps both parts of the problem: it continually removes the deposited tungsten; not only that, it returns the tungsten to the filament, thus increasing the life. Other effects come along with the change to the halogen cycle, some good and some not so good. We shall see about those, after we have had a look at the working of the cycle.[1]

The vapors of the halogens iodine and bromine react reversibly with tungsten, in two temperature ranges. For discussion we first consider iodine; it is the one used in most of the halogen lamps. At a moderate temperature (a few hundred degrees C) iodine vapor will combine with tungsten metal to form WI_2. The vapor reacts in that way with the tungsten that has been deposited on the lamp's inner wall. If the wall is hotter than 250°C, the WI_2 that is formed does not stick, but goes off as a gas. It then diffuses, and when it encounters the much hotter filament it dissociates into $W + 2I$. The W stays on the filament, but the I atoms do not; they diffuse away and eventually get back to the wall to again react with W. So the cycle is one of cleaning the wall and returning the tungsten to the filament. Since the iodine works over and over, only a trace of it is necessary. The lamp contains, in larger amount, an inert gas, usually argon. In some special applications, particularly photographic, bromine vapor is used instead of iodine vapor. Iodine gives a slight red-violet color to the light, which sometimes is not wanted. Bromine affects the color much less.

The relatively complicated and costly physical design of halogen lamps is due mainly to the requirement that the inner wall of the envelope be hot (in practice, between 250 and 1200°C). The envelope has to be of quartz and of small diameter: quartz because of its high melting point (1650°C) and small diameter (~ 1 cm) so it will be heated to sufficiently high temperature by the filament. Because of the small size (smaller than the end-joint of your little finger) and the high temperature, the mounting and electrical lead-ins are fragile. The lay user is not trusted to make those connections: the quartz part is sold already installed in the parabolic reflector or other enclosure (Fig. 1, a and b). The enclosure may be sealed (as in the headlight) but the sealing is only for the purpose of keeping out dust and moisture so the reflector will stay bright. Higher wattage lamps, up to kilowatts, mainly for industrial use, come with connector bases on the quartz lamp itself (Fig. 1c).

Finally, how does the halogen lamp perform? First, the presence of the trace of a halogen does not change appreciably the efficiency (lumens per watt at a given filament temperature) from what it is in a conventional tungsten lamp, *at the start*. The only difference in that respect is that the efficiency holds up throughout the life, because the wall stays clean. Second, due to the return of tungsten from the wall to the filament the life, at a given temperature, is extended by about a factor 2. One might think the extension would be infinite, since all the tungsten is returned. But it does burn out, and for the same reason

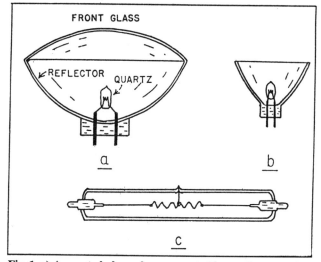

Fig. 1. a) A quartz halogen lamp mounted in a reflector, as in a headlight. b) A lamp mounted in an open reflector, used in home projectors. c) One of many styles of high power lamp.

that an ordinary filament burns out. Any filament is unstable; that is, if due to fluctuations or imperfections in manufacture it starts out with some local spots that are a little thinner than the average, the temperature and therefore the evaporation rate at those places will be greater than average, so the spots get still thinner, until one of them finally burns out. The return of tungsten from the wall by the halogen cycle does nothing to compensate the instability — that is, it doesn't fill in the thin spots. So the burnout process is not basically changed. A third point is that an excellent beam is produced by the halogen lamps that come mounted in parabolic reflectors. The source of light is small in dimensions, it is accurately mounted at the focus, and in the case of headlights, which are sealed, the reflector stays bright and clean. On the negative side we can cite only cost: for example $20 for a 300-W projector lamp, mounted in a 2-in. aluminized plastic reflector. Automobile halogen lamps cost less, about $10 at a place like K-Mart, but more at a service station.

Halogen lamps come in great variety, but mainly in high wattages, starting from the small slide-projector lamps of a few hundred watts to big ones of 5 kW or so, for purposes like theater stage or airport lighting. The spectral distribution is another variable. About 3000 to 3100°C filament temperature is normal, for a life of the order of 1000 hours. But they are made, for photographic and other purposes, with temperatures up to 3400°C, with correspondingly shorter life, and a spectral distribution moved toward the blue.

References
1. Information is from GTE-Sylvania technical bulletins, kindly supplied by David A. Scioli.

Locating the Studs in a Wall

How many times have you thumped a wall thoroughly with your knuckle, then driven a nail, only to have it miss the stud? Help has arrived. Being a frustrated stud-misser, I answered an ad for an electronic stud-sensor. I admit I ordered it partly to see what was inside it. You may not be a do-it-yourselfer who needs to know where a stud is, but even if not you may find the solution interesting. On a test where I knew the location of the studs, the gadget gave all the right answers. It is a little bigger than a hand-held calculator — 2½ x 5½ x 1 in. (Fig. 1). The indicator is a column of five light-emitting diodes (LED's). To find a stud, you hold the device with its back flat against the wall for a moment, with the on-button pressed. After some flashing for a few seconds, the LED's settle down, all of them dark. In that process it has calibrated itself for "no stud." Of course, if you happen to start out holding it where there *is* a stud, its self-calibration will be defeated. Try again in another place. With that done, you slide it along the wall surface, and when it goes over a stud some or all of the LED's light, starting from the bottom of the column. You go back and forth to locate the stud accurately.

The lid of the device was removable, but a look under it was not enlightening: a circuit board printed on both sides, integrated-circuit (IC) chips with unfamiliar numbers, etc. So other nondestructive tests had to be thought up and tried. The advertising flyer says the device is *not* a nail detector. That made it pointless to look for simple magnetic effects. The remaining possibility was that there might be a high frequency field near the device, of such a frequency that it could be tuned in on a radio receiver. So I held the device near my ham receiver, and there got a surprise. Not only was there a signal, but it tuned in at intervals of 7 kHz, up and down the whole range of the receiver. That is just what one would expect if the device were generating, not a sine wave, but a series of sharp

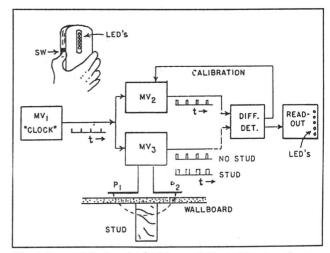

Fig. 1. Upper left, the physical form of the stud locator. Center, block diagram of the electronic units. The train of pulses from each MV is indicated. The feedback loop that makes the initial calibration is shown at the top. The capacitor plates, with an electric field line traversing the stud are shown schematically; actually their shape is a little more complicated.

pulses, 7000 of them per second. (Fourier analysis would show that.) But no matter what the nature of the pulses, I expected that if I would bring a block of wood up to the back of the device, something about the received signal would change — most likely the 7 kHz spacing. But — second surprise — that made no change whatever, of any kind that the receiver could discern. At that point I gave in and wrote to the company.[1] Bob Franklin, the co-inventor and designer, was most accommodating. So I can give you the gist of how the thing really works. It strikes me as being very clever.

We had best start by considering the stud, in the schematic diagram in Fig. 1. The property of the stud that comes into play to make its presence known is its *dielectric constant*. Just beneath the surface of the back of the stud locator is a capacitor made of a pair of metal plates, widely enough separated so that the electric field between them reaches out and goes through the stud, as suggested by the dashed line. When a stud is present, its dielectric constant (higher than that of air) causes a small increase in the capacitance of the pair of plates.

To see how the change in capacitance of the pair of plates is translated into the blinking of the LED's, we now go to the beginning of the circuit. It starts with three multivibrators (MV's). (If you are not familiar with MV's you may want to read Reference 2, before going further.) MV_1 is free-running: it makes 7000 pulses (called "clock pulses") per second. MV_2 and MV_3 are one-shot; they are triggered simultaneously, 7000 times per second by the clock pulses. The capacitor consisting of the two large plates P_1 and P_2, is connected into MV_3 so that a change in its capacitance, due to the presence of a stud, will change the *length* of the output pulses of MV_3. That change in the pulses is indicated at the right of MV_3. During the calibration stage (the first few seconds after the device is switched on), the length of the pulses from MV_2 is automatically adjusted to match the length of those from MV_3. After that, the MV_2 pulses serve as a standard of comparison: when a stud is passed over, the pulse lengths cease to match. The difference detector recognizes the mismatch and causes the LED's to light. The greater the mismatch, the greater the number of LED's that light. That proportional feature enables the user to go back and forth and maximize the response, locating the stud quite accurately.

The answer to one of the puzzling observations made with the ham receiver should now be evident. The separation of the signals along the frequency scale (7 kHz) is fixed by MV_1 and has nothing to do with the presence or absence of a stud. In my test the pulse length did change, but the receiver could give no indication of it.

The calibration feature evidently is needed, to take account of different thicknesses and kinds of wall board and/or plaster. To go through its circuit would not be worth while. Having it automatic seemed to me like frosting — calibration with a small knob would be just as quick. But then I grew up with knobs and dials. Not all people like them!

References

1. Zircon International, Inc., 475 Vandell Way, Campbell, CA, 95008. The device is now appearing in various mail-order catalogues.
2. A multivibrator (MV) used to be wired up out of a couple of transistors, some capacitors and resistors; now it is purchased as a chip. It can be made to operate in the *one-shot* or the *free-running* mode. In the one-shot mode, when triggered by an input pulse, it produces one output pulse and waits for the next trigger pulse. The shape (especially the length) of the output pulse is determined by the resistors and capacitors of the MV, not by the height or length of the trigger pulse. It has terminals to which an external capacitor and/or resistor can be connected, to make the length of the output pulse any value desired. In Fig. 1, MV_2 and MV_3 are one-shot. The other mode in which an MV can be operated is free-running. MV_1 in the figure is free-running. It requires no external trigger. It triggers itself, putting out a succession of pulses. The repetition rate can be made any value desired (in our case 7 kHz), by an externally attached resistor and capacitor.

How to Lift a Ship without Doing Much Work

Last summer, after watching ships being lifted up and down 65 ft, I had the delayed reaction of not understanding the changes — or exchanges — of potential energy. It looked as if the cost in potential energy was zero. After seeking out and reading some descriptive literature on the particular locks, I could again believe the laws of physics. A clever scheme, you will agree.

The locks referred to are the famous hydraulic lift locks at Peterborough, Ontario, on the Trent-Severn Waterway, started in 1896 and completed in 1904. They raise ships 65 ft in a single stage, still the world's record. If you imagine a large bathtub full of water mounted on top of the hydraulic lift your service station uses when greasing your car, you have the basic idea. The version for lifting ships is sketched in Fig. 1a. First of all, there must be two lifts, moving in opposite phase (see the end view in Fig. 1b). They are approximately balanced against each other at all times, because of the water pipe connecting the two hydraulic rams (the two vertical cylinders with pistons). Are

they put out of balance because one or the other of the tanks (called boat chambers) contains a ship? No, provided the water level in the tank is not changed, because the ship displaces just its own weight of water. What about the overall change in potential energy? No again. When a ship enters the tank at the lower stop, its weight of water is forced out. When it leaves at the upper stop, its weight of water moves in, to be carried down on the next trip. Ship and water have been exchanged, with no overall change in potential energy. When the ship returns, the exchange is in reverse. A closed cycle.

What puzzled me (in later thinking) was that the system operated silently — no sound of motors or pumps. Yet it was plain to see that there were large friction forces; that they would require energy to be supplied from somewhere. The solution is one we would not think of — yet invented before 1896. In all the time I spent watching the locks operate, *I failed to see it*: the key to the energy problem. It is that the tanks stop about a foot short of matching

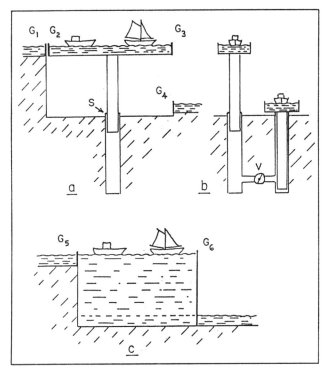

Fig. 1. a. Side view of one of the boat tanks in the up position. G_1 and G_4 are the gates at fixed height; G_2 and G_3 are the gates on the tank. At S there is a water-tight packing gland surrounding the piston.

b. End view of the two lifts, one up, one down. V is the valve that lets one lift go down and the other up, at a controlled speed.

In the foregoing sketches, it goes without saying that there is a lot of structural bracing, not shown. There are fixed vertical rails, and rollers on the chambers, to guide the chambers when moving and keep them horizontal.

c. What a conventional type lock would look like if it were to raise boats the same distance, 65 ft. Gate G_5 would offer no problem, but G_6, 65 + 8 ft high, would be subjected to enormous force. Enough water to change the level 65 ft would be expended each trip.

the inside and outside water levels. Before the gates are opened to let the ships in or out, the water levels are equalized (through a bypass that is not visible). So to start the next trip, the water in the upper tank is 2 ft deeper than that in the lower tank — more than enough imbalance to take care of friction forces. Then all that has to be done to make one tank go up and the other down is to let water flow from one hydraulic ram to the other. Of course many further mechanical features and safeguards could be mentioned, but the foregoing has given the essential physics.

Some dimensions of the Peterborough lift are of interest, since it is the highest in the world. The tanks are 140 x 33 ft, and 8 ft deep. The rather shallow depth indicates that the main traffic is pleasure boats, house boats, etc. Often a dozen are loaded at once. The hydraulic rams are 7½ ft in diameter, and the water pressure in them is nearly 600 psi. The connecting pipe with the control valve in the middle is 1 ft in diameter.

Lift locks are used the world over, although not to the extent that the conventional type locks are used. They have the advantage where the flow of water in the river or canal is limited, or where a large change in height in a single stage is required. On the first point it is interesting to see what the water requirement would be if the Peterborough lock were of the conventional type and still a single stage of 65 ft (Fig. 1c). To raise a ship, the whole compartment 140 x 33 x 65 ft would have to be filled with water, from the river above. That would be 33 times the volume of water used for one trip of the hydraulic lift lock, the latter being only a layer 140 x 33 x 2 ft. The other point to note is that the gate at the downstream end (G_6 in the diagram) would have to be 65 + 8 ft in height; it would be subjected to enormous force, and would be very expensive. Of course conventional type locks are not made in such great height. Where the total change in height is great the job is done in several easy stages, in series.

A visit to the Peterborough lift lock is recommended. You will spend a long time watching it work.

Reference

1. An interesting folder giving data and diagrams was received free upon request to Peterborough Kawartha Tourism & Convention Bureau, 135 George St. N., Peterborough, Ontario, Canada K9J 3G6.

More on the Thermobile: Power from Wire that Remembers

Ed McNeil in a communication to the DOING physics section (*TPT*, Sept. 1984) introduced readers to a novel kind of heat engine. Soon afterward it showed up in the Edmund[1] catalog, so clearly it deserves continued discussion in these columns. I have been helped by Frederick Wang, President of Innovative Technology International,

Inc.,[2] the company that made the "toy" engine that first caught my interest.

As Ed McNeil explained, the device depends on a wire made of an alloy that has quite a special kind of memory; it remembers, and tries to return to, the shape in which it was originally manufactured and heat treated, whenever its

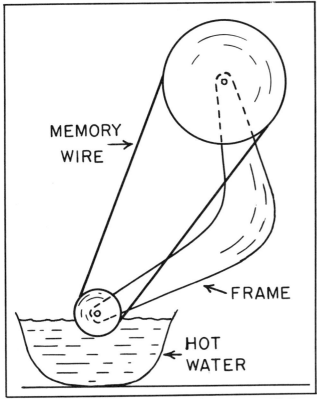

Fig. 1. Schematic diagram of the toy heat engine "Thermobile."

temperature is raised above a certain threshold. For example, a piece of the wire, manufactured straight, may be quite flexible at room temperature (23°C) but, upon being heated to 50°C, will straighten itself, exerting some force in doing so. That property makes possible the little heat engine shown in Fig. 1. A piece of such wire (manufactured straight) is in the form of a loop, around two pulleys. The lower pulley (about an inch in diameter) is partially immersed in a cup of hot water. The engine must be given a start, which can be in either direction. The power that runs it comes from the strong internal straightening force in the

wire, as it *leaves* the small pulley. It has just passed through the hot water, and has been heated above its threhold for "shape memory" (and the shape it remembers is straight). That power is not balanced off by power required to bend the wire as it goes *onto* the small pulley, because there the wire is flexible, having cooled below its shape-memory threshold during its travel through the air, over the large pulley.

It may not be easy to see how the tendency of the wire to straighten makes the pulley rotate — the configuration is unfamiliar. We might think, instead, of a spool with some (originally straight) spring wire wound on it. If you hold the end of the wire and release the spool, the spool will rotate, and be able to do work.

The wire of the engine is made of an alloy called Nitinol, named for the components and the place of origin: **Ni**ckel **Ti**tanium **Na**val **O**rdance **L**aboratory.[3] That, and a number of other alloys having shape memory have been studied for a long time. The metallic alloys return to the original shape when the temperature is *raised*. More recently, some nonmetallic materials (polymers) have been found, which do the opposite: return when the temperature is *lowered*. Some history and the metallurgical and solid-state details of shape-memory materials are given in a recent article in *Physics Today*.[4] In that article, as well as in the literature from Innovative Technology International, designs for more powerful engines (for instance, employing multiple wire loops) are shown and their possible applications are discussed. But it takes a lot of wire to make a little power. Dr. Wang tells me that a 30-loop engine, made with 0.012-in. diam wire, produces 2.4 W. The same engine is being run on a life test, and is approaching 10^9 cycles, with no change. The top speed is about 700 rpm. Engines that work between room temperature and ice temperature are now being offered.[5] They do not, as one might guess, use the polymer having reverse characteristics, but use a metal wire whose transition temperature lies between zero degrees and ambient.

From a Reader: A Puzzle about Rear-View Mirrors

Why is it that when you *tilt* the mirror, by turning the little button or moving the lever, you do not change the *direction* of view (as you would expect from tilting) but change only the *intensity* of the image? Dennis Kehoe of Luke M. Powers Catholic High School (Flint, MI)[6] writes that he and his class have had much discussion trying to figure that out. From inspecting one of the mirrors, Dennis and his students found out a couple of facts to use as starting points: the glass is wedge shaped, thicker at the top than at the bottom, and it is silvered only on the back surface. They also had the observation, put in by some of

the students, that in some conditions of light the interior of the car roof can be seen in the mirror. No combination of internal reflections seemed to account for the observation. In the end the class decided to send me a mirror and put the problem in my lap.

The key to the solution, which apparently Dennis and his class did not make use of, is the fact that the *unsilvered* front surface of a piece of glass will show an image: a dim one, only about 5% of the intensity of the image shown by a silvered surface. In many applications of rear-silvered mirrors the weak image in the front (unsilvered) surface is a

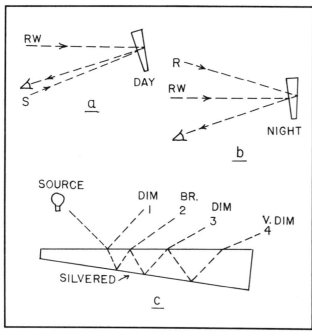

Fig. 2. a. Daytime. Light that comes through the car's rear window (RW) reaches the driver's eye by reflection from the back (silvered) surface of the mirror. Light from the seat (S) or person in it, reaches the eye by front surface reflection, too dim to be noticed.

b. Nighttime. The mirror has been tilted. Light through the rear window (RW) from headlights behind reaches the eye by reflection from the front (unsilvered) surface. Light from the interior roof (R) reaches the eye by reflection from the back (silvered) surface. But the roof is nearly dark.

c. By holding the mirror near the eye and looking along paths 1, 2, 3, and 4, bright, dim and very dim images of the light bulb are seen. Both internal and external reflections from the unsilvered surface are involved.

nuisance. But in the rear-view mirror, it is turned to practical use. An image reduced to about 5% is just what is needed for viewing the glaring headlights of cars following behind at night. The lines of sight for the two positions of the mirror are shown in Fig. 2, a and b. Although, in fact, two different images are present in both cases, one of them, depending on the light conditions of night or day, predominates and the other is not noticed. In daytime (a) the image of the brightly lighted outdoor scene completely overrides the faint image of the front seat, or of the driver. At night (b) the images of the headlights behind (even though reduced to 5%) easily override the image of the dark interior of the car roof. (But of course if you turn on the light inside the car, you may see the roof.)

After receiving my reply, Dennis and his students did further experiments. By holding the wedge shaped mirror close to the eye and viewing the images of a bright object (e.g., an electric light) they could identify a further order of image, one of intensity reduced by two reflections at unsilvered surfaces (about 5% of 5%). Their sketch is reproduced in Fig. 2c.

References

1. Edmund Scientific Co., 101 Gloucester Pike, Barrington, NJ 08007.
2. 10747-3 Tucker St., Beltsville, MD 20705.
3. Frederick E. Wang, Bernard F. DeSavage, and William Buehler, Jour. Applied Phys. 39, 2165 (1968).
4. Ahmad A. Golestaneh, Phys. Today 37, 62 (1984).
5. "Cool-Craft" and "Icemobile," by Innovative Technology.
6. 2040 W. Carpenter Rd., Flint, MI 48505.

Stretch Orientation: A Process of a Hundred Uses

If you are in a canoe, without a paddle, on a river that is flowing toward a gap between two walls of rock, going faster and faster, relax and have no fear that the canoe will arrive at the gap crosswise. The reason is that, since the water is speeding up as the stream narrows, the water at the leading end of the canoe moves faster than the water at the trailing end. (Assuming of course that the canoe didn't start out exactly crosswise.) The effect is to turn the canoe toward an orientation parallel to the flow of the water. The same happens to any sticks that may be floating in the stream. The phenomenon is called *stretch orientation*, and the term applies whether the material concerned is a liquid or a deformable solid like a rubber band. It just means that the material lengthens while its cross section area decreases. The orientation occurs for objects all the way from canoe-size down to molecule-size, provided

they are long in shape. Figure 1 shows how it goes for long chain molecules.

Assuming you don't have the problem of being in a canoe without a paddle, let us shift to the smaller scales of size, which are more interesting. You may have found, in removing an article from a newspaper, that tearing the paper up or down goes straight and easy; but not so, across the page. That is because the wood fibers, of which the paper is made, lie predominantly up and down the page, the direction in which the paper came off the roll when printed. Rolls of paper are made in a continuous process, in which a soup of wood fibers suspended in water flows out of a slit onto a moving belt of screen (screen to allow the water to drain out). The solidified but still wet layer then gets peeled off and passed through hot drying and pressing rollers. The soup, in flowing out of the vat through the narrow slit is stretched, and that aligns the fibers parallel to the direction of stretching. So, the paper ends up having a "grain."

The story of how Polaroid (light-polarizing plastic sheet made by the Polaroid Corporation) got started is interesting and to the point. In the form that first was widely used (called J-sheet) the polarizing was done by microscopic-size needle-shaped particles of a dichroic[1] crystal, herapathite, embedded in the plastic. The polarizing ability of herapathite had been known for a long time[2,3] before Edwin Land, the young inventor who later founded the Polaroid Corporation, became interested in it. But the substance was not thought to be of any practical use, because crystals of it were fragile, and could not be grown even large enough to cover the eyepiece of a microscope without great difficulty. Land's winning idea, in about 1926, was to seek ways to orient millions of the microscopic needle-crystals parallel, and thus have them act as a single crystal. After some tries (to some degree successful) at orienting them by electric and magnetic fields, he hit upon the simple way: embedding them in a plastic and stretching the plastic (while soft).

In time, a still simpler way of making the polarizing sheet was found; possibly suggested by the fact that the addition of iodine had been the key to making the dichroic material in Herapath's experiment with dog's urine. It was found that if a plastic (polyvinyl alcohol), whose long polymer molecules had been aligned parallel by stretching, was dipped in an iodine solution (like the tincture found in the medicine chest) the plastic became strongly dichroic! The parallel molecules of the plastic formed the framework for the orderly arrangement of the iodine atoms. Polarizing sheet made in that way became the H-series, the type in most common use today.

One more facet of stretch-orientation, before we stop. That is the fact that certain liquids, consisting of randomly oriented long molecules, will become *solid* if the molecules are made parallel by stretching. It happens because the parallel arrangement makes more bonding possible between molecules. Sugar, a long molecule, behaves in that way. Concentrated sugar syrup can be made solid by stirring or, as Grandmother used to do, pulling it to make taffy.

The most intriguing way in which the above effect seems to play a role is in the spinning of webs by many kinds of insects. The best studies of the physics and chemistry of the web-making process, including the role played by stretch-orientation, have related to the making of silk by silkworms.[4] Assuming the process is basically the same for all web-spinners, it will be interesting to think of an example we often see in action: spinning by spiders. Web is known to consist largely of long chain molecules. The material in the gland, inside the insect, has to be fluid — a tangle of the long molecules. But *on the way out* through a fine duct and nozzle, it has to acquire enough tensile strength to support the insect, who may be descending rapidly by hanging from it and spinning (Fig. 2). In the passage, the molecules get oriented parallel (as in Fig. 1) which adds strength to the web. Further items: Most

web-spinning insects have not a single nozzle, but one or more clusters (called *spinnerets*), each nozzle supplied by a gland. The individual nozzles are exceedingly fine, giving a very great stretch-factor. Nature has not overlooked one other important matter: the web contains a sticky component, so spiderwebs can catch flies, webs strands can be attached together, and cocoons can be constructed.

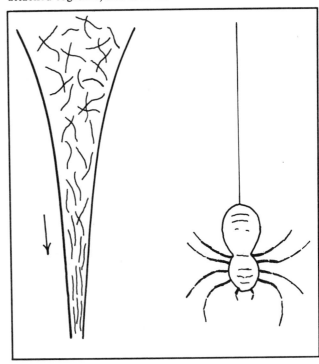

Fig. 1. (left) How parallel orientation occurs when a substance containing, or comprised of, long molecules or particles is forced through a narrowing channel or slot.
Fig. 2. (right) "Look, Ma! No hands!" Some spiders at times descend, rapidly, hanging from the web and spinning.

References and notes

1. Dichroic materials have the property that they absorb light of one plane of polarization and transmit (are transparent to) light of the perpendicular plane of polarization (or component thereof). They do not rotate the plane; they just remove one component. The term dichroic comes from the fact that the property, in some of the materials, is wavelength-dependent, so in white light they show colors.
2. In Edwin Land's words from J. Opt. Soc. Am. **41**, 957 (1951), "In the literature there are a few pertinent high spots in the development of polarizers, particularly the work of William Bird Herapath, a physician in Bristol, England, whose pupil, a Mr. Phelps, had found that when he dropped iodine into the urine of a dog that had been fed quinine, little scintillating green crystals formed in the reaction liquid. Phelps went to his teacher, and Herapath then did something which I think was curious under the circumstances; he looked at the crystals under a microscope and noticed that in some places they were light where they overlapped and in some places they were dark. He was shrewd enough to recognize that here was a remarkable phenomenon, a new polarizing material.
   "Doctor Herapath spent about ten years trying to grow these green crystals large enough to be useful in covering the eyepiece of a microscope. He did get a few, but they were extremely thin and fragile . . . ." Note: Herapathite is quinine sulphate periodide.
3. William Bird Herapath, Phil. Mag. (4th ser.) **3**, 161 (1852).
4. *Encyclopedia of Polymer Sci. and Technology* V. 12, p. 579 (Wiley, New York, 1970).

Measuring Alcohol Concentration in the Blood

Testing for the concentration of ethyl alcohol in the blood of a vehicle driver is a routine and quick operation that can be performed by a police patrolperson. It is done by a measurement of the alcohol concentration in a sample of exhaled breath. The fact that it gives a fairly reliable result depends on a number of controls and "human-proof" procedures that have been worked out over the years, relating to both the collection of the sample and the chemical analysis. We shall look into both of these; first the sample collection.

One might question whether the alcohol concentration in the breath is a reliable indication of that in the blood, considering that breathing may be fast or slow, deep or shallow. We know, according to *Henry's law* that if the liquid (blood) and the gas (breath) are kept in contact long enough for *equilibrium* to be established, there will be a definite ratio of the concentrations of alcohol in the two phases. (We of course have to note that "contact" is not direct; it is through the membrane that lines the lungs, which is still quite permeable to the alcohol.) The question is, how nearly does the breath sample have the equilibrium concentration?

The requirement is surprisingly well satisfied, if a simple procedure is followed in the collection. The enormous surface area within the lungs over which air is exposed to the blood capillaries causes equilibrium to be reached quickly. If a subject holds their breath for a short time and if only the last (the bottom) half or third of the air exhaled is collected, the sample will have quite accurately the equilibrium concentration. It goes without saying that imbibing must not have occurred recently enough that the vapor comes from the mouth. The collection of the sample is where the care and judgement of the officer are most important — the remainder of the procedure is pretty much programmed, as we shall see.

In almost all of the devices the alcohol is measured by letting it reduce a strong oxidizing agent — one that is colored to begin with and which loses its color as it is reduced. The loss of color then is the measure of the amount of alcohol. The favorite oxidizers are potassium permanganate and potassium dichromate. Water solutions of these are purple and yellow, respectively, and colorless when reduced. The breath sample is bubbled through one or the other of these solutions, and the reduction in color measured. In designing devices to do this, the aim is to *eliminate the possibility of any variation* in the process, as different officers use it and under different conditions. Succeeding paragraphs indicate how this requirement is met.

A device called the "Breathalyzer,"[1] which is used widely and is official for the State of Michigan, will serve

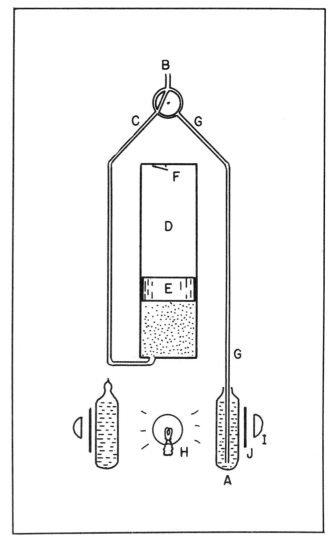

Fig. 1. Schematic diagram of the Breathalyzer. The parts are identified in the text. The dots below the piston indicate alcohol vapor. (The driver did not pass the test.)

for our description. It will show the basic ideas used in most devices.[2] We start with Fig. 1. Before starting the test, a new glass ampule (A) containing a standardized solution of potassium dichromate is installed so that the gas tube (G) dips to the bottom. (The ampule comes sealed, and the top has to be broken off.) The breath (the bottom third exhaled) is introduced at B, with the three-way valve set so it flows to C. (The breath may be introduced directly or after it has first been collected in a balloon and brought to the central station.) The gas flows into the holding cylinder (D), pushing up the piston (E). When the piston gets to the top it signals the fact by closing an electrical switch (F). Next the three-way valve is turned to connect C to G, and the gas starts bubbling up through the solution in the ampule, pushed by the weight of the piston. The rate of bubbling is standard, determined by the weight of the piston and the size of the hole in the tube — *it is not under the control of the officer.* Likewise, the amount of gas, and the amount and concentration of the solution in the ampule are standard.

Next comes the determination of the degree of color loss by the oxidizing solution. This is done photoelectrically. There is an incandescent lamp (H) between the live ampule and a sealed, comparison one. There is a photoelectric cell (I) on either side, each behind a blue filter. (The solution is yellow by transmitted light because it absorbs the blue; therefore the change is best measured in the transmitted blue.) The electronic circuitry compares the responses of the two photocells, and gives a reading on a voltmeter whose scale has been changed (empirically, no doubt) to read directly the percent of alcohol concentration in the blood. When the device is used on the road, power for the electronics and for keeping the oxidizer in a prescribed temperature range, can come from the car battery.

Breath-alcohol testers have been around since about 1934. In the last half-century, numerous inventors and manufacturers have entered the competition[3]; some have disappeared. And the literature has been voluminous — not a little of it purporting to show that particular (or all) of the methods are untrustworthy. The evolution has resulted in a few surviving devices that are simple to operate and that are considered by the police and judges to be reliable. Different states have slightly different definitions of what percentage of alcohol in the blood is cause for arrest. In Michigan the dividing line is 0.1%. (An amount of liquor containing two ounces of ethyl alcohol, which is consumed by a 150-lb. person and allowed time to reach the blood stream, produces this concentration.)

References

1. The Breathalyzer is made by Smith and Wesson Electronics Corp. of Eatontown, NJ.
2. For the description of apparatus and procedures, I have received much help from Sergeant Arthur Hughes of the Ann Arbor Police Department and from Lyle Filkins of the University's Transportation Research Institute.
3. A survey of devices available at the time (1972) is given in a pamphlet "Breath-Alcohol Tests" published by the American Medical Association.

What Is the Secret of "Maintenance-Free" Car Batteries?

I had always believed you had to add distilled water to the battery at least as often as oil to the crankcase. Then within about the last decade, there quietly came on the scene batteries having no filling caps, thus requiring no additional water for their lifetime. Impossible, I thought, unless a radical change in principle had been made. So I made up my own theory. Everybody knows (or knew) that in the charging process some electrolysis always goes on, producing gas (H_2 and O_2) and depleting the water. Had they had put a catalyst inside the battery to recombine the hydrogen and oxygen? Fine theory, but dead wrong!.

Inquiries to some battery companies[1] brought interesting results. There are no changes *in principles,* from the traditional batteries, the kind with caps. There is no catalyst. They are not absolutely sealed; there is a small vent you can find if you look hard enough. The reduction of water loss to near zero has come from the combination of two separate improvements: one in the batteries themselves, and the other in the electrical system that charges them. First the batteries.

The active, rechargeable materials in a "lead-acid" battery, when charged, are lead dioxide (PbO_2) on the positive side and lead (P_B) on the negative side. The electrolyte is diluted sulfuric acid (H_2SO_4). In discharging, both the Pb and the PbO_2 are converted to lead sulfate ($PbSO_4$). The reverse occurs during charging. These reversible reactions do not in themselves result in the decomposition of water into gas. But the conditions are not ideal: in the practical battery other things go on that do decompose water. The so-called grid makes most of the trouble. The active lead compounds are porous - spongy - and so have to be supported by a metal framework, the grid. During charging, not all of the current goes to the sponge; some goes to the grid, and there it decomposes water into gas. Battery makers have made much progress in reducing this cause of water loss by developing new alloys for the grids, and by reducing their surface exposure to the electrolyte.

Improvements in battery construction are only half the solution to reducing water loss. Any battery, improved or not, will make gas when the charging current is excessive (exceeding the rate at which the sponge can react), or when charging current is continued after the battery is fully charged. Both of these causes of water loss have been all but eliminated by radical changes in the charging equipment, thanks mainly to the advent of solid state components. So due to the two attacks, battery caps are disappearing.

Just possibly your car may be one which has a battery with caps and a non-solid state charging system; therefore it will be interesting to see what the differences are. In the older cars the generator is the DC type, and charging is regulated by an ingenious and complicated system of relays[2]. After you've seen them described, you may find it hard to believe they have made millions of cars run as well as they have! There are three relays in the circuit of Fig. 1. They are shown as they are when the engine is not running: the contacts of R1 and R2 closed, and those of R3 open. R1 regulates the voltage. Its coil is in parallel with the output of the generator, so when the generator voltage exceeds a predetermined value, its contacts open.

The contacts are in series with the field winding of the generator, so when they open, the generator voltage starts dropping towards zero. Before it drops far, the contacts close again. The process repeats many times a second, keeping the voltage within a narrow range (ideally a little above the 12.6 V of the fully charged battery). But that is determined just by the *tension of the spring* in the relay. It is set at the factory, but very likely adjusted later by a service person who has the philosophy that too much is better than too little. Result: gassing and water loss.

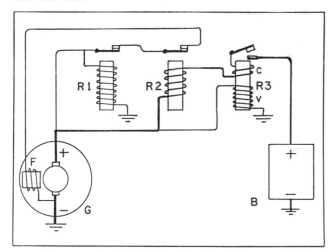

Fig. 1. G, the car DC generator; R1, R2, and R3, the control relays, and B, the battery. R1 and R2 control the charging voltage and current by intermittently opening and closing the circuit of the field coil F of the generator. R3 prevents discharge of the battery back through the generator. Heavy lines indicate the path of the charging current.

Relay R2 works much as R1 does - interrupting the field circuit of the generator and regulating its voltage. But its coil is in series in the charging line, so it lowers the generator voltage when the *current* goes too high. It does its duty when the battery is low, or dead. In that case R1 would still let the voltage go to 12.6 or above, causing excessive current, making gas, losing water, and even damaging the sponge. R2 prevents that.

As if this were not enough relays, there is one more, R3, a very important one. Its contacts are in series with the charging line;

its purpose is to prevent the battery from discharging back through the generator when the engine is idling or stopped. It has *two* coils: a "voltage coil" (V in Fig. 1) and a "current coil" (C). Together they give a "latching" or "hold-down" effect. Coil V closes the relay when the generator voltage is high enough for normal charging (above 12.6). If charging takes place, as expected, the current in coil C is in the direction to add force to holding the relay closed, even if R2 lowers the voltage due to the battery being low. If for any reason the current flows in the wrong direction, the coil acts in oppposition, and contacts open.

The modern system is simpler. The generator is an AC one: an "alternator." Gone is the multi-bar commutator and the brushes that wear out. DC is made by rectification - silicon diodes. Gone also is the problem of the battery discharging through the generator. The current can't go backwards through the diodes. Protection against too-high a charging current is not needed; the alternator is designed so it will not give excessive current. Finally, the voltage is regulated by a solid state circuit (now a chip) which acts by varying the field current of the alternator. The standard of comparison for the voltage regulation is a Zener diode.[3]

The Zener diode is highly accurate and stable. There is even correction for temperature. The sensor for that is a thermistor.[4]

If you are to be left with one message, it might be that if yours is a chugger with a DC generator, you should not let your service person sell you a maintenance free battery. Unless you want to drill holes in the top for adding water!

References

1. The best information was supplied by Lynn Wik, of GNB Batteries Inc., P.O. Box 43140, St. Paul, MN 55164. It included the *Technical Battery Service Manual,* by Battery Council International, 111 E. Wacker Dr., Chicago, IL 60601.
2. W. E. Billiet and L. F. Goings, *Automotive Electrical Systems,* 3rd Ed., 1970. Published by American Technical Society, Chicago, IL.
3. Zener diode: a specially concocted diode which, when operated under reverse voltage, begins conduction very sharply at a particular voltage, say 12 V. Zeners can be made with any desired "breakdown" voltage. They are used for precise reference voltage in regulators.
4. Thermistor: a ceramic semiconductor, that has a very steep resistivity-temperature relation, in a restricted temperature range. By varying the composition, the sensitive temperature range can be made to suit the purpose. Useful for temperature controllers or temperature references. See *Phys. Teach.,* p. 60 (1983).

Fluorescent Lights: A Few Basic Facts

Antoine du Bourg, of the Pingry School[1] sent me the results of experiments he had made to find out how fluorescent lights work, and encouraged me to write about them. That got me started doing some "reverse engineering" of my own. Finally, to fill in the many gaps in our findings, I appealed to the GTE-Sylvania Company, and was helped by David Krailo.[2]

There are three principal questions that can be asked: (1) how is the electrical power converted to light, (2) what is the function of the "ballast", and (3) how is the gaseous conduction in the tube made to start—or "strike"—when the switch is turned on? Answers to the first two can be made in common to all the types of fluorescent fixtures. The answer to the last question, how the conduction is started,

has a lot of variations. We had best start with the features that are common. First, the way light is produced.

Unlike the way light is produced in the familiar neon sign, very little (only about 2%) of the electric power applied to the gas column is converted *directly* into visible light. About 60% is converted to *ultraviolet* radiation; the remaining 38% is converted to heat. The ultraviolet is absorbed by a thin layer of fluorescent material on the inner surface of the tube. This re-radiates the energy in the visible wavelength range.[3]

The conversion from ultraviolet to visible is nearly 40% efficient. So overall, 60% x 40% gives a remarkable 24% efficiency for the process, about twice that as for a tungsten filament lamp.

How Things Work

The ultraviolet radiation comes from mercury atoms. The tube contains a noble gas, e.g. argon, with just a small amount of mercury vapor. It is at the optimum low pressure for gaseous conduction. The current flows along the gas column mainly by ionizing the noble gas. But the mercury vapor does almost all of the radiating. Its principal spectral line is at 253.7 nm wavelength.

The *ballast* is, in its simplest form, a coil wound on a laminated iron core: an inductor; a "choke." It weighs a pound or two, gets warm, and sometimes hums. Its main function is to stabilize the gaseous conduction. Why the tube would be unstable without it may be explained as follows. Conduction by ions in a low pressure gas has the characteristic that if the current increases, the ionization density increases, which *lowers* the resistance. So if the voltage applied to the tube is *fixed*, the lowering of the resistance increases the current, which lowers the resistance still more, sending the tube along a path to self-destruction. All that is necessary to stabilize it is to insert in series either a resistor or a choke, so that when the current increases, the voltage across the gas tube will decrease. The choke is preferred over the resistor, because it wastes much less power by converting it to heat. In some fixtures the "ballast" may be more than a choke; it may have transformer action as well, to increase the voltage above the line voltage.

It is in the ways of starting the gaseous conduction that we find the wide variation of schemes. The problem in all cases is, in short, that the voltage necessary to keep the gaseous conduction going is not enough to start it. The ways of initiating this action fall into four classes: *preheat, rapid start, instant start,* and *trigger start.*

The preheat system is the most common. The basic circuit is shown in Fig. 1. The initial ionization to start the gaseous conduction is produced by electrons from filaments at either end of the tube which are heated momentarily. The interesting question is how the filaments are switched on automatically, for only as long as needed. The gadget for doing that is called, in the engineering labs, a *glow-switch,* and at the hardware store a starter. It is like the familiar small (¼" or ½" diameter) neon light, except that one of the electrodes inside is a bimetal strip. When gas conduction occurs, the bimetal gets hot, bends and makes contact with the other electrode (see Fig. 2). This puts the current through the filaments of the fluorescent tube, and of course it extinguishes the glow in the glow-switch. In the second or two it takes the bimetal to cool and re-open the contact, the filaments have come up to temperature. When the contacts open, the full voltage[4] is across the fluorescent tube and, with the help of the electrons from the filaments, it starts. The glow-switch does not restart because, when the fluorescent tube is conducting, the voltage across it is less than the threshhold of the glow-switch.

The *rapid start* scheme differs from the preheat in that a glow-switch is not used; the filaments start heating when the main power

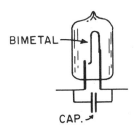

Fig. 2. Enlarged sketch of a glow-switch. Some (but not all) have a capacitor as shown, presumably to reduce arcing as the contacts open.

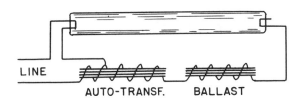

Fig. 3. Simple, one-tube instant start circuit. The tube shown has the standard two pins; they are short-circuited inside.

switch is turned on, and their heating current continues while the fluorescent tube is running. It deserves the designation rapid start because it does not have the delay during which, in the other type, the glow-switch heats up and closes the filament circuit. Evidently, the filaments are so designed that running them continuously doesn't unduly shorten the life of the tube.

In the *instant start* scheme, heated filaments are not used. The "ballast," instead of being just an inductor as in the other systems, includes a step-up auto-transformer that produces a voltage much higher than the line voltage. It is enough to start conduction in the tube without the help of a filament. The simplest possible instant start circuit is shown in Fig. 3. It is interesting to see the use that is now made of the *two* pins at the end of the tube, which in the other systems were necessary for heating the filament. Here they are used as a safety measure. The high voltage from the step-up transformer would present a shock hazard, if a tube were not in place, leaving the sockets exposed. This hazard is avoided neatly: the two pins of the tube are connected together inside, and the line circuit goes in one and out the other. No tube, no power.

You may be reading this under a fluorescent desk light. If so, it probably is of the *trigger start* variety. This system is used commonly for the small (15" or 18" wide) tubes. There is no glow-switch. There is a red and a black button. You heat the filaments momentarily by pressing the red one. This also puts the line or ballast voltage on the tube. The black button turns off the line voltage.

The several types of circuits have been presented in simplified— perhaps overly simplified—versions. There are more complex circuits, mainly in the *two-tube* fixtures which are very common. Space in this column will not permit going into detail. One feature is called "lead-lag." The two tubes are phased so that their peak light outputs come at different times, thus smoothing out the flicker. Another is called "sequential starting." It applies to the instant start system. The tubes are operated in series, but for starting, *all* of the voltage is put on one and then on the other.

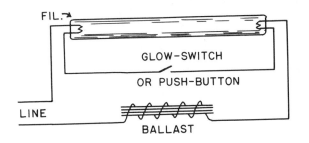

Fig. 1. Basic, one-tube preheat circuit. The filaments are heated, momentarily, either by a glow-switch or a push-button.

Clearly, there is quite a field for experimentation. Antoine and I recommend it as a pastime!

References

1. Martinsville Road, Martinsville, NJ 08836.
2. Lighting Products Group, 100 Endicott St., Danvers, MA 01923.
3. The fluorescent compounds are often called *phosphors*. One can be found that will fluoresce in any color of choice, from red to blue. A desired spectral distribution, for example daylight, can be made with mixtures.

Examples of compounds are Cadmium Borate (pink); Calcium Halophosphate, Silicate and Tungstate (white, orange, blue, respectively); Magnesium Tungstate and Germanate (red, bluish-white respectively); Strontium Halophosphate, light green; Zinc Silicate, green. There is even one, called 360BL that fluoresces in the near ultraviolet, for "black light" tubes (from Sylvania Engineering Bul. 0-341).

4. After reading the draft Antoine du Bourg pointed out, logically, that the tube may get a surge of voltage higher than the line voltage, due to collapse of the magnetic field in the ballast when the glow-switch opens. That would vary, of course, depending on the instant of opening within the AC cycle.

The Air Bag: An Exercise in Newton's Laws

It's a challenging problem: how to bring a human body to a stop, without disastrous injury and in 3 ft or less, from 30 mph. Advocates of the air bag ("inflatable restraint system") claim they have the most reliable means for doing this, and thereby saving countless lives. Great controversy goes on as to whether air bags should be required by law, whether they will add excessively to the cost of a car, and so on. Bypassing all that, we physicists may find some fun and surprises in examining the problem as a case of the laws of uniformly accelerated motion—or should we say *decelerated* motion? No, we will be proper and say accelerated, keeping in mind that the numbers will be negative. Some experimental data will be needed, to put into the equations; mostly the kind found from car crash tests. Help in finding that information has been given generously by Dr. John W. Melvin, of the University of Michigan Transportation Research Institute.

It may be helpful first off to see how the air bag works and what it may be expected to do or not do. The air bag is a kind of balloon, packed into a small box that is installed in the center of the steering wheel. When a collision occurs, the bag inflates and forms an air-pillow between the driver's chest and head, and the steering wheel. The inflation must be, and is, very quick. The gas, although called air, is nitrogen. About 2 cu ft of it is produced by the explosive decomposition of sodium azide. Electric switches at the front end of the car and at other locations are closed by sudden acceleration of -10 g or more. This puts current through an electric detonator cap in a capsule of sodium azide, making it explode.[1] The bag is fully inflated about .04 s after the front of the car hits an obstacle. That may seem like extremely fast action, which it is, but it is none too fast, as we shall see when we begin putting the numbers into the equations.

The air bag can do only one job: *spread* the force over the person's chest and head. That alone increases the chance of survival. In designing the whole system employing the air bag, the automobile companies have paid attention to at least two additional factors that can reduce the damage to the person:

1. The distance over which the acceleration occurs should be made as great as possible. The greater the distance, the smaller the force, for a given change in velocity.
2. Uneven, high peaks of force on the person, which would be especially damaging, should be avoided. That means striving to make the (negative) acceleration of the person as near uniform as possible.

The main source of greater distance in which to bring the person to rest may surprise you; it is from the crumpling of the front end of the car. You trade off your own damage against damage to the car! This varies considerably among makes and models of cars. In some planning for "crash-worthiness," greater distance of crumple has been designed into the front structure. The second source of distance, one that is advocated as part of the safety "package," is a steering column

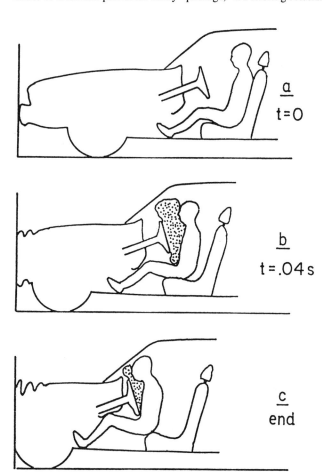

Fig. 1. The three important stages in the crash of an air bag equipped car against an immovable barrier.
(a) The car has just made contact with the barrier.
(b) The air bag has just become fully inflated.
(c) End of the collision: all parts have just come to rest.

that will telescope under the appropriate large force. That can give an additional six or eight inches. The third source of distance is the deflation of the air bag. At full inflation it may be as much as a foot in thickness, filling the space between the person and the steering wheel. If, by a fairly large intentional leak, it decreases to a few inches during the acceleration process, the person gets some additional distance to go. Of course, the sum of the distances gained from the steering column collapse and the deflation must be not quite enough to let the head smack into the windshield! The foregoing acts only on the upper torso. To complete the "package," a deformable panel is mounted under the dash which pushes against the knees while deforming about six inches.

The hard part for the design engineers, and what probably is the important part of the crash tests, is to make:

1. the resistance to crumple of the front end,
2. the resistance to telescoping of the steering column, and
3. the leak of the bag, such that the person is accelerated uniformly, and brought to a stop just as the full available distance is used up.

We have to assume that ideal performance in order to be able to make our simple calculations. Therefore, our result will be a "best case."

We can get the few numbers we need from the results of tests—the kind using instrumented dummies. They seem to be standardized: 30 mph against a concrete barrier, equivalent to the collision of two equal mass cars travelling at 30 mph each. We will work with this, which (switching now to mks units) is the same as 13.41 m/s. For the crumple of the front end, .70 m seems to be a good average from the tests. The time for inflation of the bag seems to be quite definite, .04 s. For the initial chest-to-steering wheel distance, .36 m is reasonable.

Figure 1 shows three stages of the collision:

1. the first contact of the bumper with the wall,
2. the completion of the inflation of the bag (.04 s),
3. the end of the collision.

We do a few calculations—not shown, because the equations of uniform acceleration are entirely familiar. First, from the numbers just given, we get the acceleration of the car during the crumple. It is -13.1 g. In the .04 s interval before the bag is inflated, the person continues to move forward at the initial 13.41 m/s while the car is slowing down at -13.1 g, so the chest-to-steering wheel distance decreases to .26 m. In the same interval, the front end collapses .43 m, leaving .27 m to go. This situation is shown in Fig. 1b. From Fig 1b. to 1c., the bag deflates. We take .08 m for the final thickness which gives a change of .18 m (.26-.08). The steering column telescopes .20 m and the front end collapses its remaining amount, .27 m. Adding up, we have a total of .65 m in which the person must be brought from full velocity, 13.41 m/s, to rest. That requires an acceleration of -14.1 g.

If the person's upper torso has a mass of 40 kg, the force is 5540 N, which is over half a ton in the English system. The lower torso gets even more, because of the shorter distance the knee panel deforms.

As mentioned at the beginning, all of the calculations are made using the simple equations for uniformly accelerated motion, which, of course, are at your fingertips. If you find it interesting, or if the car you drive happens to be a compact, you might think about how the calcuations would have to be revised for a head-on collision of your compact with a full-size car, each initially travelling at 30 mph. Think of the standard unequal-mass putty-ball collision: the combined putty-balls do not end up at rest. You would have to make your own assumption as to the distances of crunch of the two front ends-they certainly would not be equal. What force would each driver have to withstand? Another thing you might consider is the time element. I found, for the conditions I used, that the times at which the front end crunch ended and the air bag/steering column collapse ended, differed by only .008 s. Very close; that is, all parts of the system came to rest very nearly at the same instant. I would not bet on that being true for cars of unequal mass. You will think of other things to check!

References

1. During the past few years a simplified version has been developed by the Breed Corp. of Lincoln Park, NJ. It is self contained, inside a box that is to be installed in the steering wheel. The sodium azide is made to explode much as a rifle cartridge is fired. It has on it a detonator cap, which will be struck by a spring-operated firing pin. The firing pin is held "cocked" by a trigger that has considerble mass. When a collision occurs, the inertia of the trigger causes it to release the firing pin. The advantage claimed, as compared to the electrical system, is chiefly that of cost. A number of cars so equipped are on the road, in a Government-operated test.

New Life for an Old Device: The Ring Interferometer

It sometimes happens that an old device or method has a rebirth because of new materials or components that have become available from neighboring fields. That happened to the *ring interferometer*, in a striking way. What does it do? If the entire apparatus is mounted on a turntable, and the turntable is rotated, it will measure the rate of turning with respect to the cosmic inertial reference frame; that is, with respect to the universe of stars. Let's see what the old, basic idea is, and then see how modern improvements have increased its precision by many orders of magnitude.

In Fig. la, light from a source, S, goes to a "beam splitter" (a mirror that is so thinly silvered that it reflects only about half the light intensity and lets the other half pass through). The part that is reflected goes *clockwise* around a closed circuit and back to the beam splitter. (It is made to follow the closed circuit by a series of mirrors, M_1, M_2, M_3.) On arrival back at the splitter, again part is reflected and part goes through. The part that goes through is wasted, but the part that is reflected goes to a photographic plate, P. (There is a lens, which in effect makes a camera.)

Now what about the part of the light from the source that goes *through* the splitter? It goes *counterclockwise* around the circuit, and on encountering the splitter for the second time, the part that goes *through*, goes to the photographic plate. There it combines with—interferes with—the light that has come via the clockwise path. If the two components are in phase, there will be brightness; and if they are out of phase, there will be darkness. Actually, because the geometry will not be perfect, there will be light and dark fringes on the plate which will move when the phase difference changes.

If the apparatus is at rest, the two-path lengths are the same. But now see what happens if the *whole* apparatus is built on a turntable and is set into clockwise rotation. The clockwise light has to travel a little further to get back to the splitter (and photographic plate), because during the travel time of the light, the splitter has moved to the position indicated by the dotted line. (Because the velocity of light is so great, the movement will be extremely small—only the distance of a fraction of the wavelength of the light.) Likewise, the counterclockwise light has to travel a little shorter distance. The result is that the rotation of the table causes a shift in the relative phases of the two components on arrival at the splitter. The same phase shift carries through to the photographic plate, because the splitter-to-plate distance is fixed. Light will change to dark, or fringes will move.

In 1913, the French physicist G. Sagnac[1] devised the experiment basically in the form just described and got a result. Only, he was on the wrong track in his interpretation: he presented it as a proof of the existence of an ether.[2] But the experiment was sound, and the result reproducible; so the origin of the method is still credited to him, even though the interpretation is now different. In Sagnac's experiment the path around the closed circuit was only about 3 m. The time taken for light to make that circuit is so short that he had to rotate the apparatus at 120 rpm to get a measurable effect. In succeeding years, the Sagnac method was refined so that much smaller speeds of rotation could be measured. But in very recent times a great leap in its sensitivity has come about thanks to technical developments in related areas. First is the optical fiber, a hair-thin strand of glass, plastic, or a composite that can pipe light with very low loss. That property, in combination with the high light intensity input afforded by laser technology, makes possible the piping of light for distances of a great many kilometers. Already the fibers, in fact bundles of them, are replacing wires for the transmission of information—between people and between computers. The second is the technique of sensing extremely small phase differences of light by electronic means—far smaller than what can be sensed by photography or by the eye.

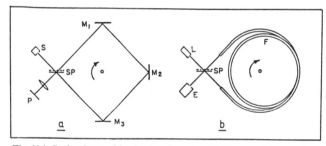

Fig. 1(a). Basic scheme of the ring interferometer. Light source S, beam splitter SP, mirrors M_1, M_2, M_3, photographic plate or film P. The splitter divides the light and sends it both ways around the circuit, to reach the photographic plate, where the two components make an interference pattern. Sagnac mounted the entire apparatus on a turntable and rotated it at 120 rpm. (b). The modern version: laser light source L, beam splitter SP, many-turn optical fiber F, and electronic phase-difference detector E.

The Sagnac type apparatus, transformed by the employment of the above techniques, and now called the "fiber optics gyroscope"[3] is shown in Fig. 1b. Light from a laser, L, is conducted around the closed circuit by an optical fiber, F.[4] Instead of going around once, enough turns are wound around to make the path about 1 km. That greatly increases the transit times for the two light paths, allowing the apparatus to turn further in that time. The photographic plate is replaced by an electronic phase-difference detector, E. With these improvements put together, the enthusiasts are now talking about achieving a sensitivity of a few times 10^{-7} deg/s. There is a difference between sensitivity and a quantitative measurement. But if we concede them only 10^{-5} deg/s for a measurement, the device should give a big reading for the rotation of the earth, which is more than 4×10^{-3} deg/s.

Applications have not been mentioned, but they are easy to think of, especially if the gyro device is made in a compact form. Navigation and guidance are the first to come to mind. On a more academic plane, there is the prediction from general relativity that the earth's motion "drags" the inertial frame with it locally, just a little, and the question as to whether that can be detected. Recently[5] an analysis has been made as to the possibility (still far away) of detecting it by means of a Foucault pendulum. Perhaps now the fiber optics gyro is in the running also.

References

1. G. Sagnac, *Comptes Rendus* **157**, 708, and 1410 (1913); *Jour. de Physique* **4**, 177 (1914).
2. Early physicists had a hard time getting used to the idea that light could be propagated through nothing—a vacuum, so they imagined that all space was filled with an "ether," and devised experiments to detect it. Sagnac's contention was that when his apparatus was rotated, an "ether wind" was produced, against one of the light beam directions and with the other, causing a slight difference in velocity.
3. B. Culshaw, *Jour. Physics* E **15**, 390 (1982); *Snippets* **4**, p 9 (1983); **5**, p 7 (1983/84). (*Snippets* is a newsletter published by the Institute of Physics, London.)
4. The fiber must be "single mode." That means it must be fine enough, and of such other properties that the propagation is a "waveguide" type, having a unique velocity. See W.B. Allan. *Fibre Optics: Theory and Practice*, (Plenum Publishing Co., London and New York, 1973).
5. V.B. Braginsky, A.G. Polnarev and Kip Thorne, *Phys. Rev. Letters*, **53**, 863 (1984).

How a Cable Car Turns a Corner

Beginning when I was "knee-high to a grasshopper," I was fascinated with the San Francisco cable cars. I could hear, and even see, the cable moving beneath the slot in the street midway between the rails, and also see the metal piece projecting from under the car down through the slot, which evidently clamped onto the cable to make the car go. Much later it began to dawn on me that there were some basic things about the system that seemed impossible.

Topologically impossible! Specific examples: how could cars get across the intersection of two car-lines, where surely one cable would have to cross under the other? Then there was the matter of the hundreds of little pulley wheels that supported the cable and kept it located correctly under the slot. Along a level street or at the crest of a hill the pulleys would have to be under the cable; in a valley they would have to be above, to hold the cable down; on curves they would have to be in the horizontal plane on the inner side of the curving cable. How could the piece connecting the car to the cable get by wheels that might be below, above, or on either side of the cable?

No more can be done here than to try to tell how a few of the mechanical puzzles are solved, leaving much unsaid.[1] First, just a remark as to the historical setting. The first cable railway was built in San Francisco in 1873. It was about a mile long and the cable was powered by a steam engine. That seemed to be a solution the world had been waiting for; in the next two decades nearly all the sizable cities in the Western world installed cable car systems. Then in the following two decades nearly all of them disappeared. It was the story of the displacement of the (rail) horse car by the cable car, and then the displacement of the cable car by the electric trolley car. A precious few of the railways survived—just because the citizens—and tourists—loved them.[2] San Francisco's seem now to be safe forever. Nothing in that railway has changed except for the replacement of the steam engine that pulled the cable by an electric motor, and the replacement of the gas headlights by battery powered ones. The original bell is the last thing one would dare to replace.

So how about the mechanical puzzles? First let's understand that the cable, an endless loop, runs continuously at 9½ mph. The car moves by seizing the cable in a "grip"; essentially a pincer that extends down through the slot in the street to the cable. It is opened and closed by a large lever in the car above. Figure 1 is a photograph of the grip. In Fig. 2**a, b** and **c** it is shown in its three conditions: closed tightly on the cable, the car being pulled; loose, the cable running freely through while the car is stopped for passengers; open, cable dropped onto the guide pulley. The grip cannot be moved up or down; it is at a fixed depth, a little *above* the pulleys. Along straight level track, the grip holds the cable up off each pulley as it passes over (Fig. 2**d**).

Fig. 1. View from below of the most important part of the cable car: the "grip" that clamps onto the moving cable to make the car go. The cable runs on pulleys just below the grip; it is not shown in this picture. (Photo courtesy of San Francisco Public Utilities Commission.)

Where the track is curved, the pulleys are in the horizontal plane, as in Fig. 2**e**. Here they are at a little smaller radius than the slot, so the grip holds the cable out, away from each pulley as it passes.

The problem gets more difficult where the street is concave upward—between a downhill and an uphill. There the pulleys have to be on top of the cable, to hold it down. A simple and clever device makes the pulley move out of the way while the grip passes (Fig. 2**f**). In those places the pulleys are located so that when the cable is running on them, it is *higher* than the jaws of the grip. So when the grip approaches a pulley it depresses the cable to a level lower than the pulley, freeing the pulley. The pulley is then moved aside, so the grip can pass. That happens because the pulley is mounted on a pivoted plate, as shown in the figure. After the grip has passed, the plate and pulley come back into place (presumably by a spring) and the cable comes up to run on it again. The problem of the intersection is solved simply, but not without its hazards (Fig. 2**g**). One cable (#1) goes through normally. The other (#2) has to go under #1. As the car approaches on cable #2, it must drop the cable and depend on momentum (or gravity if there is a slope) to take it through to where it can engage the cable on the other side. Since the grip is fixed and cannot reach down to get the cable, a *dip* is put into the track, to lower the car momentarily so the grip can close on the cable. The gripman (as the operator is called) has to be alert and accurate in timing. So, as far ahead as possible, warnings are set up. At the right place to drop the cable there is a bright-colored plate in the middle of the tracks that says LET GO. If that doesn't do it, a little farther along the grip hits a lever (below the tracks) that rings a bell. If both warnings are unheeded, the cable will be pulled forcibly out of the grip, usually tearing some of the strands. The gripman must report this immediately—not a pleasant duty!

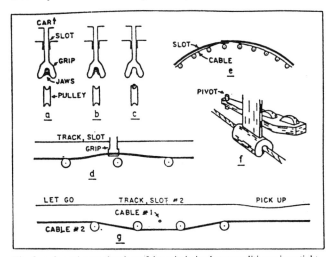

Fig. 2. a, b, and **c** are sketches of the grip in its three conditions: jaws tight, car moving; jaws relaxed but not open, car standing; jaws open, cable has been dropped onto the pulleys. **Fig. 2d.** How the grip holds the cable up off the pulley as it passes over. **Fig. 2e.** Curved track. How the pulleys are located with respect to the slot and grip so that the grip holds the cable horizontally away from each pulley as it passes. **Fig. 2f.** What happens when the track is concave upward. The grip holds the cable down, below the pulleys, and pushes each pulley (or pair of pulleys) to the side as it passes. **Fig. 2g.** The intersection of two cable car tracks. #1 goes through normally. #2 has to "let go," coast through, and pick up the cable on the other side. To enable it to pick up the cable, the track takes a dip.

Not the least of the gripman's concern is anticipating the signals and breaks in the traffic, so the car can go through without having to apply brakes, thereby losing momentum and getting stranded with no source of power. (We suppose the passengers could get out and push!)

Finally, the turn-around is the source of entertainment for countless sidewalk superintendents. The one at the foot of Powell Street is the famous one, pictured on postcards.[3] Two parallel tracks end at a turntable. The cable, moving toward the turn-table under one track and away from it under the other, reverses by going around an underground pulley whose diameter is equal to the distance between the slots of the tracks. The car must LET GO (of the cable) just before it gets to the turn-table, and roll on by momentum. The car and turn-table are rotated by human muscle (which is the main event for the onlookers). When the car has been pointed outward, on straight track, the cable, just under the grip, is pushed up by a lever (hand operated) so it can be engaged by the grip, and away the car goes.

References

1. There is no lack of literature. An especially good, short account, emphasis on the mechanics, is by James L. Delkin, *Anatomy of the San Francisco Cable Car* (Hooper Printing Co., 1946). A comprehensive account is by George W. Hilton, *The Cable Car in America* profusely illustrated, (Howell-North Books, Berkeley, CA 1971), 484 pp. A folio of detailed drawings and description of the San Francisco system "Cable Car System Engineering History" was recorded in 1981 under the Cable Railway Recording Project, assisted by federal funds. "A Brief History of the San Francisco Cable Cars" (8 pp.) is available from San Francisco Municipal Railway, 949 Presidio Ave., San Francisco, CA 94115.
2. After being badly wrecked by the earthquake and fire of 1906, the parts of the cable system involving the steep hills were rebuilt. The other lines were converted to electric cars. In the 1940's the preservationists battled to save the remaining cable lines from being replaced by bus lines, and won narrowly. By 1982 the system had nearly worn out mechanically. A huge sum, mainly from federal "National Landmark" funds was then spent restoring it, with great care not to change its character in any way. Operation is expensive. For the present reduced system, 2 Mw of electric power, 24 hr/d runs the cables; only 10% of it ends up moving the cars. The 10 miles of 1¼ in. steel cable wears out every 100 to 200 days. Grip jaws have short lives.
3. That line uses single-headed cars, which have to be rotated 180°. Other lines in the San Francisco system use two-headed cars, that can reverse without rotation.

A Plywood Goose that Seems to Fly

Every so often I come across a device, that ought to be simple-minded to understand, which defeats me. Even if not for long, it bugs me in the meantime. This was done to me recently by a wing-flapping plywood goose, found in a novelty store in Ottowa, Canada.

The goose is (of course) a Canada goose (Fig. 1). It is made entirely of 1/8 in. thick model builders' plywood, with a wingspread of 28 1/2 in. and a body length of 19 in. The width of the wings at the body is 6 in., starting from 9½ in. from the tip of the beak. The crosspiece, a 1/4 in. dowel, is 10 in. long and 12 in. above the wings. (I try, always, to have my measuring tape handy!) The dimensions are given here, because after reading the rest of the description, you probably will want to make yourself a goose. The wings are hinged to the body, so they can "flap" with very little friction. The bird

is supported from above by a system of nylon monofilament. *To take a photograph* that would show the nylon strands, white string was stuck onto them in parallel.

When the bird is at rest, the wings are horizontal. If you give the knob that hangs below a pull and then let go, the wings flap for a period of 2½ s, for quite a time. During the motion, the points of attachment of the supporting filaments to the wings stay at a fixed level; the body goes down as the wingtips go up and vice versa. It's a lifelike imitation of flight, even if too slow. Until I made a copy of the bird for myself and played with it, the mechanics of it was a puzzle: from where do the forces come that restore the wings to the horizontal position, and what are the parameters by which the flapping frequency is made to approximate that of a goose? The creator of the bird either knew physics pretty well or did a lot of cut and try! (Unfortunately, we can't give credit—there was no signature on the bird.)

Let's identify some forces—first when the bird is at rest, wings horizontal as in the photograph. To support the body, each wing must exert, through the hinges, an upward force equal to half the weight of the body. The four filaments that support the whole bird are attached to the wings, all at the same distance out from the body. That distance is such, that if a wing with half the body weight added at the end were placed on a knife-edge, defined by the two points of support, it would balance. That distance having been determined, the filaments are attached to the crosspiece above at such separation that they are vertical when the wings are horizontal.

Fig. 1. The goose in its resting position. White strings were added along the nylon monofilament strands, just to make them visible in the photograph.

When the conditions are fulfilled, the wings will be horizontal when at rest and will *seek* horizontal if displaced from it. The question is, why is the horizontal position stable, and how do the restoring forces come about? In Figure 1a. and 1b., the wings are shown displaced to upward and downward angles, respectively. In both cases, the points of attachment of the filaments to the wings on the two sides move closer together, so that the filaments are no longer vertical. Their forces then have horizontal components, indicated by p. Those must, of course, be counterbalanced by equal and opposite horizontal force components, q, exerted on the wings by the hinges, p and q are not co-linear, so they constitute a couple; a torque. That produces rotation of the wing, in the sense indicated by r, namely toward the resting position. Although not shown, all forces are mirrored on the other side of the bird, so there is no net effect on the whole bird.

It is easy to see why the restoring torque reverses, from a to b: nothing changes except that p goes from above q to below.

The conditions for oscillating motion are present: mass that can be accelerated, and restoring force (in this case torque) that acts toward the resting position. To go further, and calculate the frequency, would get a little complicated, because there is a combination of linear motion (the body) and rotation (the wings). One can, at least, see that changing the length of the supporting filaments should be a direct way of changing the frequency. In fact, I have tried it.

For persons who might want to make a goose, a few tips are offered. The hinges, in the bird in the store, were made of the nylon monofilament. I found it easier to make them like half-chain links, out of #18 brass or copper wire, pinched into #60 drill holes in the plywood. In fastening the supporting filaments to the wings, etc., it is very hard to tie knots in the nylon and come out with the right lengths. The monofilament is put through a #60 drill hole in the plywood, then a tiny U-shape clamp, made of the aluminum from an old radio chassis is pinched onto the end of it.

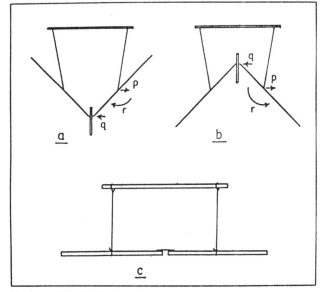

Fig. 2a. and 2b.: The horizontal components of force, indicated by p and q, produce torques that rotate the wings toward their resting positions, from either the up or the down position. Fig. 2c: A way to experiment with the system by means of three sticks, string, and tape.

For persons having no need for a goose, the interesting behavior of the system can be enjoyed with nothing more than three sticks and some string. A piece of tape can make a flexible hinge (Fig. 2c). The strings are attached securely at the cg's of the "wings" (which may or may not have loads at the ends simulating a body) and are movable at the upper stick. Things to try: if the separation of the strings at the crosspiece is decreased, there will be *two* stable resting positions, both such as to make the strings vertical. Perhaps, the designer of the goose started with three sticks.

Some of the Physics of Cardiac Pacemakers

Nearly a million persons wear pacemakers beneath their skins. The subject of pacemakers is so big, especially on the medical side, that you will have to trust me to pick out a few points that should interest physicists. I have been helped by J. R. Buysman of Medtronic, Inc.,[1] and Craig Hassler of the Battelle Columbus Laboratories,[2] both of whom wrote extensive answers to my questions, and by Susan Plaster of Cardiac Pacemakers, Inc.,[3] who supplied descriptive material.

To see what job the pacemaker has to do, we should look at the electrical and plumbing system supplied by nature (Fig.1). As we know, the blood circulates through the lungs to pick up oxygen, and then through the muscles and other organs to release it, where it produces energy. On the return trip to the lungs, it carries carbon dioxide, to be released into the breath. The blood system, of course, carries the traffic of all the other substances that have to get from one organ to another. Generally, those things can stand some interruption. But the transport of oxygen cannot be interrupted for more than a few minutes without dire consequences, so keeping that going is the continuous job of the pacemaker.

The first interesting fact is that the oxygen gathering and delivery circuits are separate, each with its own pump. That can be seen by following the arrows in Fig. 1. The right side of the heart pumps blood out into and through the lungs. The blood returns to the left side of the heart, which pumps it out through the other parts of the body. It returns to the right side, to start the cycle over.

Nature's electrical system provides the regularly-timed electrical impulses that make the muscles of the pumps contract. The impulse that starts each contraction cycle is supplied by a remarkable little piece of tissue in the top of the heart, the sino-atrial node—*SA node* for short. In modern terms, it is a "microprocessor." It senses, in various ways, the body's current and expected need for oxygen and adjusts the rate at which it sends the impulses accordingly. The response to *expected* need is a most interesting ability. Emotion (for example, due to sensing danger) results in inputs to the SA node that cause it to increase the pulse rate, so you will be prepared to fight—or run!

Now a little about the pumps. The pump in each side of the heart is a two-stage one. The upper chamber (the *atrium*) fills with incoming blood. On receiving the electric impulse from the SA node, it contracts and expels the blood into the lower chamber (the *ventricle*). Of course, there is a one-way valve between the chambers, to make the blood go in the right direction when the contraction occurs. *After* the blood is transferred to the ventricle (which takes one or two tenths of a second), the ventricle contracts and pushes the blood out to the system. That important delay is introduced by another "microprocessor," the *AV* (atrioventricular) node. It senses the electrical activity of the atrium, introduces the delay, and sends an electric impulse to the ventricle. The action that has been described applies to both sides of the heart. The left and right upper chambers are so interconnected through nerve fibers and tissue that they contract simultaneously, even if only one side receives a stimulus; the same is true of the left and right lower chambers. The same is *not* true between the upper and lower chambers, for good reason: the lower ones must not contract along with the upper ones, but must wait the .1 to .2 sec for the blood to transfer, then contract upon receiving the delayed signal from the AV node.

We begin to see what the job of the artificial pacemaker is.[4] Essentially, it is to send electric stimuli to replace those normally furnished by either the SA or the AV node, or both. To make the problem more complicated, in most cases of an ailing heart the nodes still do their jobs part, or most of the time, so help is needed only now and then. Early pacemakers simply sent their electric impulses regularly whether needed or not. That could confuse the heart rhythm, besides running the pacemaker's battery down needlessly. Modern ones are of the "noncompetitive" or "demand" type. That means they are able to *sense* the electrical signs of heart action and *not* send impulses in if the contractions are occurring normally.

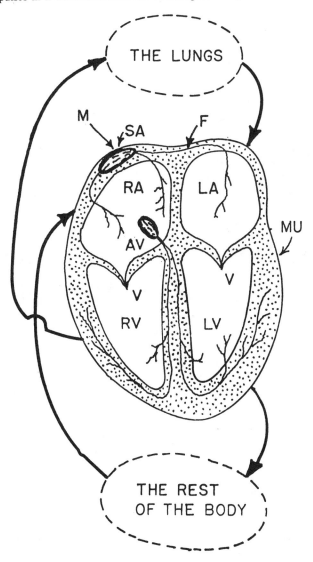

Fig. 1. Cross section of the heart of a person facing the reader, with "plumbing and electrical wiring." Only the features necessary for this discussion are shown. The letters, starting from the highest, indicate: M, messages from the body that can modify the pulse rate; SA, sino-atrial node; F, one of the many cardiac fibers carrying the electric impulses to stimulate the heart muscles; RA and LA, right and left atria; AV, atrioventricular node; V and V, one-way valves; MU, heart muscles; RV and LV, right and left ventricles. Heavy lines with arrows indicate the external routes of the blood.

How does the demand feature work? In addition to being able to send a regularly-spaced series of stimuli to the heart, this pacemaker *receives* an electrical signal when the heart contracts. (It receives it via the same wire that conducts the stimulating impulse to the heart when needed.) The circuit is such that if the received signal arrives *just before* the pacemaker is ready to send a stimulating impulse, the stimulating impulse will not be sent. Instead, the pacemaker's "clock" will be reset to zero, so it will take a full interval to get ready to send the next impulse. The waiting interval of the pacemaker is set to be a little longer than the interval of the patient's normal (resting) heartbeat; therefore, when the patient's heart is working normally, the pacemaker's clock is always reset just before it would send an impulse. So, it just stands by. If the patient's heart lags or stops, the pacemaker takes over.

There are several malfunctions of the heart that can be helped by a pacemaker. The most common is "conduction blockage." The atrium functions, but a (delayed) signal from the AV node is not received by the ventricle. In that application, the pacemaker connection is made to a ventricle, usually the right one. (As explained, that suffices for the two ventricles.) In the mode of stimulating the ventricles only, the optimum time sequence between the contractions of the upper and lower chambers is more or less lost, and there is some loss of pumping efficiency; however, that kind of demanding pace, although not perfect, is the simplest and most commonly used.

More sophisticated pacemakers and connections are used as required. In the case of conduction blockage (just described), proper sequential contractions of the upper and lower chambers can be restored by sensing the contraction of the atrium via one connection, then sending the stimulating impulse to the ventricle via another connection, with the proper delay put in by the pacemaker. In the case of failure of both of the nodes, a pacemaker that will substitute for both can be installed. A pacemaker can even be "reprogrammed" (have its timing sequence changed) by magnetic or RF signals sent to it from a device held over it *outside* the body.

Some of the physical aspects are interesting. The pacemaker is usually implanted under the skin of the upper chest, and the insulated wire lead is passed from there through a blood vessel into the ventricle—or the atrium. The batteries are of several varieties. Many are the lithium type, akin to the battery in your watch, but larger. In "demand" service, the life may be upwards of five years. Small batteries and solid-state circuitry on a chip make the pacemaker small—shaped like a cookie, typically 4 x 5 x 1 cm.

References

1. Medtronic, Inc., 7000 Central Ave., N. E., Minneapolis, MN 55432.
2. 505 King Ave., Columbus, OH 43201.
3. P.O. Box 43079, St. Paul, MN 55164.
4. A general description in layman's language: *Medtronic Currents: An Overview of Pacing*. Available from Medtronic, Inc., 1979, 77 pp.

A Spinning Coffee Can and More on Alcohol and Locks

This being the last "column" for the academic year, a little tying up of loose ends may be in order, especially those relating to subjects on which readers have commented.

Remember the alcohol breath tester, in the September[1] issue? At about the time I was sending that in, an interesting piece about the same subject in *The New Scientist*[2] was on the way from across the ocean. Guy Faucher, of E'cole Polytechnique in Montreal[3] wrote me some comments about it. In England, during May, 1983, the results of breath-testing machines were made admissible and binding as evidence in the courts. There was an immediate and continuing hue and cry from citizens, claiming that the machines were sending innocent people to jail. (Not everyone was on the losing side, however: one publisher quickly came out with a handbook for the use of defending lawyers, setting forth the technical arguments by which doubt might be cast upon the machine results. Five thousand copies were sold!) In response to the public criticism, the Home Office mounted an extensive field test. In over 12,000 instances of the use of the breath machine, simultaneous, or near simultaneous, analyses of the blood and/or urine for alcohol content were made. The result was that in 80% of the cases in which the breath machine would have let the person go free, the blood/urine test indicated arrest! That silenced one kind of objection. The controversy goes on; meanwhile, 600 breath machines are in daily use by the police.

The machine[4] most widely used in England works on a different principle from that described in this column[1] (used extensively in the U.S.). The English version depends on the fact that alcohol vapor strongly absorbs infrared radiation at specific wavelengths. It uses the one at 3.39 μm. The basic system is shown in Fig. 1. It has two long tubes, one for the suspect's breath, and one for a reference "breath," the latter being air that is in equilibrium with a water solution of alcohol at the just-intoxicated concentration. A rotating "chopper" (a sectored disk) alternates the IR between the two chambers. There is a filter (of the interference type) in front of the detector, which passes IR only in a narrow range around 3.39 μm. Automation is complete: the machine prints a tape, ready to be handed to the judge. (In the latter lies another objection from citizens: the idea that one can be sent to jail by a piece of tape from an impersonal machine.)

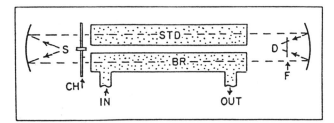

Fig. 1. The Lion "intoximeter." S is an infrared source, white-hot ceramic. Mirrors at the two ends reflect the IR through cells BR, containing the suspect's breath, and STD, containing a reference mixture of alcohol vapor and air, ending at the IR detector D. F is a filter that passes IR in a narrow range around a wavelength that is absorbed strongly by alcohol vapor. CH is a chopper (rotating sectored disk) that lets the IR go alternately through the two absorption tubes. Electronic circuitry sorts out the resulting responses of the detector.

What should be an advantage of the method has been used as an argument against it. Lung-fulls of breath can be blown into the machine repeatedly, so the tape can contain a series of results. We physicists like to have the average of a number of runs; but the objectors contend that the scatter (up to 20%) only proves that the machine is unreliable. A more real problem of the method lies in the presence of other gases in the breath (waste products of metabolism) that absorb IR around the $3.39\ \mu m$ wavelength. Acetone is the main offender, especially so because its presence varies widely among individuals. A means of detecting and subtracting its effect has been incorporated into the machine, but acetone is still believed to be the main source of error in the method.

Following the discussion of Yale locks in this column,[5] Albert Bartlett (our former president of AAPT)[6] wrote in with some questions and speculations as to how the insides of the locks can be made to accept *submaster* keys as well as master keys. For the answer, I had to go back to Gerald Huller, who is in charge of the key system of the University.

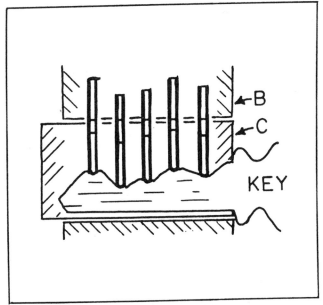

Fig. 2. The essential parts of a Yale lock. B is the barrel, the stationary part of the lock. As shown, the key has lifted the pins up to such heights that their first cuts are at the boundary of the cylinder, C. That will allow the cylinder and key to turn. A second set of cuts is shown, probably for a master key. More sets of cuts can be made, for submasters.

Figure 2 will remind us of the structure of the lock. To allow the cylinder to turn, the key must push the pins up to such heights that their cuts are at the boundary of the cylinder. For the private keys, each to open only one door, the cuts in the pins are different in every lock. To accommodate a master key, a *second* cut is made in each pin, and those cuts are the same in all the locks.

For the submaster problem, I had expected a more tricky solution than what Gerald Muller described to me. It's just done by making more cuts in the pins. Suppose all the *classrooms* are to be put on a special key, to be issued to the teachers, and perhaps a few others. A *third* set of cuts is made in the pins of the classroom locks, and corresponding keys (all the same) are issued to persons privileged to have them. Suppose the wood shop, machine shop, and the store room are to be on another special key. Since that group is separate from the classroom group, its submasters can also use a *third* cut

in the pins (different from the third cuts for the classroom locks, of course). The plot thickens if the Chairman wants to issue keys to certain persons that will unlock the classrooms and *also* the shops and store room. That calls for *fourth* cuts in the pins of all those locks, with corresponding keys.

The game can be pursued up to a total of about five cuts in each pin. But it is a "zero sum game" for two reasons. First, the heights at which cuts are made in the pins are quantized; that is, they are (typically) at 10 discreet levels. Only about half that many can be used for cuts, because it is the *un*cut parts that hold the lock from turning. Second, cuts that are used for the master and submasters are taken away from the possibilities for cuts for the private keys. The only way to expand the possibilities is to add more pins, which sometimes has to be done.

Fig. 3. The spinning coffee can, and the three pipes presumed to enclose conductors carrying 3-phase alternating current.

Finally, I pass on to you something interesting to explain, sent to me by Paul Zitzewitz of the University of Michigan-Dearborn.[7] He wrote: "While visiting Glen Canyon Dam, I saw an interesting example of what is probably an induction motor, but I'm not sure. Outside, near the large transformers, were three large pipes running vertically. I assume they carried the 3-phase alternating current output of the generators. Near them, some one had put a coffee can (empty) on a pointed support (Fig. 3.). It was spinning rapidly. Can this be explained easily for (or by) *TPT* readers — or for me and my colleagues?

References

1. *Phys Teach,* **23**, 386 (1985).
2. *The New Scientist,* July 4, 1985, p. 26.
3. Dept. of Engr. Physics, P.O. Box 6079, Sta. A, Montreal, PQ H3C 3A7, Canada.
4. Intoximeter, made by the Lion Co. in Wales.
5. *Phys Teach,* **22**, 179 (1984).
6. College of Engr. and Applied Sci., Univ. of CO, Boulder, CO 80309.
7. Dept. of Natural Sci., Univ. of MI, Dearborn, MI 48128.

Bar Codes Are on Everything—
What Do They Say?

Until I looked into it a little, at the urging of a number of readers, I had not really thought about how the price of an item comes from its bar code. The simple assumption would be that it is in the bar code. How could one think otherwise, seeing it happen so fast: the waving of the article over the slots of the reader, the beep, and the appearance of the price on the video display? But on reflection, we know that the storekeeper would not want to be bound by a price printed on the package by the manufacturer, even in a form undecipherable by the customer. And, of course, the price is not in the bar code. The scheme is clever and a little complicated; in finding out about it, I received much help from James McIntyre, of POS Cash Register and Computer Sales.[1] (POS stands for point-of-sale, what you and I call the checkout counter.)

Here's what I found out as to how the price gets into the system. The information encoded in the bars pertains only to the description of the article (8-oz can of tomato soup), the brand, the company, etc. When the article is wiped over the slot in the checkout counter, the bar code is scanned optically and converted to a series of electric signals that go to a minicomputer. The computer does an operation called "look-up." It gets, out of its memory bank, the information needed at the point of sale: the price and name of the item, for the visible display that the customer sees and for the printing of the paper sales receipt.

There are two parts to the memory bank. There is a part that contains the translation of the bar code into article description and supplier information. That part is universal (not pertaining to the local store); it is ROM (read only memory) which is in the equipment when installed. Then, there is a RAM (random access memory) part that can be changed at will by the storekeeper, via the computer keyboard. Into that is put the price, number purchased from the supplier, etc. Thus, the storekeeper can change the prices at any time; an 8-oz can of tomato soup can be described by the same group of bars the world over, yet be sold for a different price in every store. When processing each sale, the minicomputer does other useful things, such as keeping tabs on how many of each article have been sold, keeping running or daily totals on dollar sales (for the store or by checkout station), and, very important, flagging the items that are in short supply.

One cannot help being impressed, and maybe a little worried, at the nonchalance with which the checkout person waves the article over the reader. The speed

with which the package is moved varies, and little attention seems to be paid to the orientation: whether the direction of motion is at right angles to the bars, as one might think it should be, or whether the scan starts from one end of the bar group or the other. Has perhaps a bar been missed, doubling the price? No, that is no more possible than getting a wrong party on the telephone by leaving out one of the digits of the number. The computer attends to everything, leaving nothing to the checkout person: It corrects for the speed, it understands equally well whether the sequence is from one end or the other, and it verifies that the full bar group has been scanned, no bars missed. If it accepts the scan, it sends the price and other information to the display and printer, accompanied by the familiar beep. If no beep, simply try again. So even if the checkout person were blindfolded, you would have nothing to worry about. Just more tries would have to be made, to get the beep.

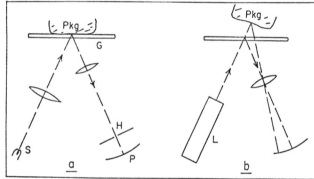

Fig. 1. (a) The essential elements of a scanner using an incandescent light source, S. In order for the photoreceptor, P, to receive light only from an area 0.3 mm in diameter on the bar code, there must be a defining hole, H, and the bar code must be in contact with the glass, G, to be in focus.

(b), A scanner using a laser light source. Since the laser beam illuminates only a 0.3 mm spot on the bar code at any distance, the defining hole is not necessary, and the package may or may not be in contact with the glass. Light paths for both cases are shown.

Improvements in equipment have been rapid, so any description will have to be qualified. Although equipment that is installed today is different from that of even five years ago, much of the earlier equipment is still in service. Typically the checkout counter has a slot covered by glass, below wich are a light source and a photocell, as shown in Fig. 1a. In earlier models, the light source is an incandescent bulb. The resolution in reading bars must be about 0.3 mm, so either the light spot or the spot that the photocell sees must be that small. That requires that the bar code be exactly at the focal distance, which can be true only if it is in contact with the glass as it passes over. If the bar code is a little above the glass when it passes over, there will be no beep. In equipment now being installed, that problem is solved by using a laser as the light source (Fig. 1b). Its beam of light can be 0.3 mm in diameter, and it can illuminate a spot that size at *any* distance. The photocell views a larger spot, so its focus is not critical. So, the necessity for keeping the bar code on the glass is relaxed. One more thing about which the checkout person doesn't have to be careful!

In early models (many in service), the computer, with its memory and its keyboard through which the storekeeper inserts prices, is separate from the printer/display register at the checkout counter—it may be located in a back room. In current equipment, the keyboard, computer, visual price display, tape printer, and in some cases the memory bank, are all in one unit at the checkout point. In a small store with a single checkout counter, the memory may be in the unit. In a multiple-terminal store, the memory may reside in one terminal, which serves as the master, and/or in a separate box. The terminals "talk to one another"; each terminal can access the memory wherever located, thus duplication is avoided. The memory capacity of modern systems is as much as 60,000 items—much more than most stores can stock. State-of-the-art equipment for a supermarket costs just under $10,000 per checkout lane.

There are many different systems of bar codes. Most of us will come in contact only with the one used for retail merchandise. That is the Universal Product Code (UPC). Examples of it are shown in Fig. 2. The most common forms are the 0-digit (a and b) and the 6-digit (c and d). In the former, the bars are separated into two groups by bars at the center that have a little extra length. A few more extra-length bars, at the left and right ends, have functions related to the operation of the system, not with the merchandise. The bars on one side of the center give all the necessary information relating to the supplier of the item. The bars on the other side give the generic description of the item, for example the 8-oz can of tomato soup. Each of the bar groups translates (in the computer) to a 5-digit number. There is no secrecy about the translation. The two 5-digit numbers are printed under the bar code on the package, as seen in the figure. In your transaction at the checkout counter, the numbers are ignored by the scanner and the computer. They are redundant, but they serve a purpose: Presumably, without a computer, you could look up the two 5-digit numbers in a master list somewhere and find that the item was an 8-oz can

of tomato soup made by the Campbell Soup Company. As an example of what has just been said, *a* and *b* in Fig. 2 show the codes for two different products made by the same company: sandwich bags and paper cups, left and right. Note that the left sides are the same.

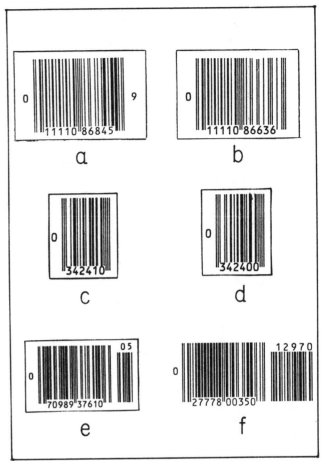

Fig. 2. (a) and (b) Bar codes from two products of the same brand, sandwich bags and paper cups, respectively. Note that the first 7 digits are the same.
  (c) and (d) Bar codes from two small Hershey's chocolate bars, with and without almonds. Note that only one of the digits is different.
  (e) and (f) Examples from paperback books.

For small items, such as 2-oz candy bars and packages of cigarettes, a single group of bars, translating into 6 numbers, is used. That requires some sacrifice of information, and of internal checks. Fig. 2c and d show 6-digit codes for two small Hershey bars, with and without almonds, respectively. Where more information than that contained in the usual 10 digits is needed, an extra 2-digit group of bars is added (Fig. 2e). In some other cases, notably paperback books (Fig. 2f), an extra 5-digit group is added. All of these are readable by the standard checkout counter system and computer.

There is remarkable agreement upon a universal code system by retail merchants. A bar code uniquely describes a certain manufacturer and product. Such agreement on a single system did not happen with phonograph records, audio and video tape recorders, television (earlier), and many other inventions.

Learning to read the bar codes is a tempting challenge. The answer to each is given in numbers below the bars, and the number of examples to work with is unlimited. That should make it a cinch for any amateur cryptographer. I have had no success. Neither have I been able to understand it from my helper, Jim McIntyre, who warns me that it is not as simple as one might think. Try it!

Reference

1. 777 S. Wagner Road, Ann Arbor, MI 48103.

Smart Ignition Systems on Modern Cars

Unless you are pretty old, you missed out on the pleasure of filing the breaker points on the chugger to stop it from "missing." Since the days when that went with owning an automobile, new and good things have happened. The breaker has gone, replaced by a transistor. New methods of obtaining accurate timing of the spark from the rotation of the engine have been devised. In the latest systems, a microprocessor gathers information—engine speed and temperature, manifold vacuum and other data—and computes continuously the optimum instant in the compression cycle for the spark to occur. Those are the new things.[1,2] The elements that have changed little are the "spark coil" that steps the 12 V of the battery up to 30 to 40 thousand, the distributor that sends the high voltage to one spark plug after another in sequence, and the spark plugs themselves. The world awaits new ideas for those—especially the spark plugs!

To be clear on the basic functions to be performed, let's look at the ignition system that has been around for more than a half century, and use a four-cylinder engine for the example. In Fig. 1, A is a spread-out schematic of the system. The cam (a) and the distributor arm (b) are on the same shaft, geared to the engine, so they go around in unison. Fig. 1B shows the same parts in their physical arrangement. When the cam pushes the contacts (c,c') open, the collapsing magnetic field in the spark coil (d) induces a surge of voltage in the secondary high enough to jump the gap in the spark plug (e) as well as the small gap between the distributor arm and the stationary metal lug (f). As the cam and distributor arm go around together, the spark occurs in the four plugs in sequence. The shaft turns at *half* engine speed, because a given cylinder is ready to fire only once every two revolutions[3] of the engine.

Because it takes a little time for the explosion to build up in a cylinder, the breaker points must open and make the spark in advance of the time at which the piston starts its downward "power stroke." There are only two ways to make that advance: Change the location of the breaker so that the cam will open it earlier, or turn the cam with respect to the shaft that comes from the engine, so that its bumps will push the breaker open earlier. Both things are done, for different reasons. We will talk of one at a time.

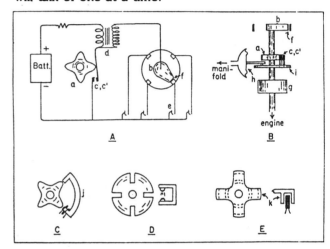

Fig. 1. (The meanings of the lower-case letters are given in the text.)
A: Schematic circuit diagram of the automobile ignition system, BE (before electronics.)
B: Physical arrangement of the mechanical parts on the shaft that is geared to the engine. From the top: distributor arm, cam and breaker points, "vacuum" diaphragm, and centrifugal device. C, D, and E: "Contactless" pickups for timing—working on magnetic, infrared and eddy current effects, respectively.

The biggest requirement for shifting the spark advance comes from the variation in engine speed. The reason is that, while the *time* of build-up of the explosion in the cylinder is roughly constant, the spark advance has to be made in terms of the *angle* of rotation of the engine, or of the shaft that carries the cam. Therefore, as the engine speed (rpm) increases, the *angular* advance has to go up about in proportion. At very high engine rpm, the advance required may be as high as 40°, with respect to the piston motion, or half that with respect to the shaft that turns the cam (which, as pointed out, turns at half the engine rpm).

The change with engine speed is achieved by a

centrifugal mechanism, g in Fig. 1B (inner structure not shown). That gives an angular displacement to the part of the shaft above it with respect to the part below it, depending on engine rpm. It shifts both the cam and the distributor arm. A simple solution to the rpm problem!

The second—and the only other variation that is put into the timing in the traditional system—takes account of the fact that the time of development of the explosion depends on the amount of fuel–air mixture that is admitted to the cylinder. If you have your foot down on the accelerator, so the pressure of the mixture is light when the spark occurs, the explosion will develop faster. The required change in the spark advance is made merely by moving the breaker points. The changes are not large enough to require shifting the distributor arm. A diaphragm (h in Fig. 1B) that moves in and out with changes in the "vacuum"[4] in the intake manifold turns the disk (i) on which the breaker is mounted.

Over the years, the replacement of the foregoing mechanical components by electronic ones has been piecemeal. The first was the transistor for opening and closing the primary circuit of the spark coil. The breaker points remained, but they then served only to control the transistor; therefore, they no longer switched the large current and no longer required maintenance. Next came the elimination of the breaker points even as a timing control and the substitution of a "contactless" control. That took at least three different forms, depending on the automobile company, all quite ingenious. They are worthy of some description.

One of the "contactless pickups" works by magnetic effects. It requires a wheel made of iron, with teeth (Fig. 1C). It is on the shaft that rotates at half engine rpm, as was the cam. A permanent magnet (j) produces flux through a coil that is wound around one end of it. As the wheel turns and the teeth pass through the position shown, there is momentarily a closed magnetic path through the iron wheel so the flux increases, inducing a voltage pulse in the coil. That voltage pulse is used for timing the spark. Another pickup for producing a timing signal works by IR (infrared light). It requires a slotted wheel on the shaft (Fig. 1D). The source is a small solid state IR-emitting diode, and the detector on the other side of the wheel is an IR-sensitive diode. When the slot passes between the diodes, a voltage pulse is produced which is used for timing the spark.

The third kind of pickup reminds us of the metal detector discussed in these columns earlier.[5] It is an inductance–capacitance circuit, driven at its resonant frequency by a transistor. A metal wheel on the shaft (Fig. 1E) has plates that pass close to both sides of the coil of the resonant circuit. Eddy currents in the plates dissipate power and cause the voltage in the resonant circuit to drop. The drop is detected, probably by a simple diode, and that is the signal used for timing the spark.

The final chapter of our story concerns the introduction of the microprocessor. We are in the midst of that change—only some of the cars have it so far. The microprocessor starts with the signals from one of the pickups we described (magnetic, infrared etc.), which give it the positions of the pistons as a function of time, and also the rpm (no more need for the centrifugal device). It gets the intake vacuum from a sensor in the manifold. And it goes much beyond: sensors in other parts of the engine give it coolant and fuel mixture temperatures, barometric pressure, humidity, and exhaust temperature. From all those data it continuously computes (or looks up from a program) the optimum spark timing, to send to the transistor that opens and closes the spark coil circuit. In that way the microprocessor can make a real improvement in the engine efficiency under varying conditions, and, partly as a consequence of the efficiency, reduce the exhaust polution.

Ignition systems for some cars (for example the Toronado) are as fully converted to electronics as possible, which means all but the distributor. The voltage from the spark coil that must be switched from one spark plug to another is normally between 30 and 40 thousand, far too high for a transistor. Distributors have been improved, but the basic principle remains.

Take a look at your ignition system, and see where it fits the foregoing descriptions!

References
1. Many questions about modern systems were answered by Frank H. Jamerson and James H. Bechtel, of the General Motors Research Laboratories, Warren, MI 48090-9055.
2. General treatments: Emmett J. Horton and W. Dale Compton, "Technological Trends in Automobiles," *Science* **225**, 587 (1984); Marvin Tepper, *Electronic Ignition Systems*, Hayden Book Co., Rochelle Park, NJ, 1977.
3. Two revolutions are necessary because each of the following requires 180 degrees: intake, compression, power stroke (explosion), and exhaust.
4. The engine, with its intake valves, acts as a vacuum pump. If the admission of the air–fuel mixture to the intake manifold is restricted, as when the motor is idling or under a light load, the suction of the motor makes a partial vacuum in the intake manifold. The degree of "vacuum" is a good indicator of what the pressure of the mixture in the cylinder will be at the instant of spark-ignition, and therefore of how fast the explosion will develop.
5. H. Richard Crane, "Metal locators and related devices," *Phys. Teach.* **22**, 38 (1984).

Beyond the Frisbee

An earlier installment of this column took up the physics of the Frisbee.[1] Now there is an interesting addition to be made. If you will look in the *Guinness Book of World Records*, you will find that the object thrown the farthest by a human arm is not the silver dollar that went across the Potomac, not a boomerang, not a Frisbee, but something new called the Aerobie.[2] Scott Zimmerman, a college student, made the record at 321.54 m (1046 ft, 11 in.). The Aerobie is thrown the way a Frisbee is thrown, and it goes about twice as far. It differs in structure in that it is a thin ring (Fig. 1) of diameters 33 cm (13 in.) outside and 25.4 cm (10 in.) inside, while the Frisbee is a disc of about 23 to 25.4 cm (9 or 10 in.) in diam, with no center hole.

Let's be reminded of what the problems are and then see how they were solved in more efficient ways in the Aerobie than in the Frisbee. The basic problem is evident, if you try to make a spinning flat disk sail through the air and stay in the horizontal plane. It slowly turns about an axis approximately parallel to the direction of flight and then falls down, edgewise. The reason, to use aeronautical terms, is that the center of lift is ahead of the center of gravity (cg). That results in a couple, or torque, that makes the disk precess, as a gyroscope. (A more detailed explanation is given in the earlier column, on the Frisbee.) The remedy is to move the center of lift rearwards to the cg. But how can this be done? Only by experimentally changing the contour of the surface. It is not hard to find a contour that will work for a given speed of flight. The real problem comes in finding a solution that stays right throughout the large range of speed during the flight.

By cut and try, maybe with the help of a wind tunnel, the Frisbee was designed to satisfy the above. And one must say it is quite remarkable: Over the range of speed, the precession is acceptably low. Evidently, the curled-down edge does it. But that produces a lot of turbulence in the air flow which, while evidently solving the range-of-speed problem, brings with it a penalty: Turbulence wastes energy, thereby increasing the drag, and shortening the flight.

The challenge is to try to find a shape that will create little or no turbulence and still satisfy the lift/cg requirement, over the range of speed. Alan Adler, a lecturer in engineering at Stanford University,[3] addressed that problem about six years ago. He succeeded, with what is now the Aerobie—but only after an intermediate stage with a model called Skyro. We can learn something from the Skyro.

Adler worked with a ring, rather than a disk. To have lift, any surface or airfoil has to impart downward velocity (momentum, properly speaking) to the air it passes through (action-reaction). To make that happen, the airfoil is given an upward tilt with respect to its direction of motion. The trouble with the simple ring is that the trailing part has to pass through air that has already been given downward velocity by the leading part, so the lift of the trailing part is reduced. Consequently, the center of lift is ahead of the cg. Adler's first solution was clever: Make the ring slightly conical, so the trailing part has a little more upward tilt with respect to the direction of motion than does the leading part. It worked as planned: The trailing part got more lift, thereby moving the center of lift rearwards to the cg. That was the Skyro. It made the *Guinness Book* at the time (1980) with a flight of 261.4 m (857 ft). But there remained a problem: The angle of the cone was just right *for only one speed through the air.*

Adler finally went to a quite different way of changing the lift of the leading part of the ring relative to the trailing part. He put a rim, that he calls a "spoiler," around the outer edge (inset in Fig. 1). The spoiler makes the airflow break away from the surfaces of the *leading* part of the wing, introducing some turbulence in the process. The result is a loss of lift for the leading part.[4] The spoiler has less effect on the trailing part, because the air gets to the spoiler *after* it has passed over its surfaces. Thus the center of lift is moved rearwards, and, if the design is right, it coincides with the cg. For reasons not easy to explain, the correction to the center of lift

Fig. 1. Photograph of the Aerobie, which is (33 cm) 13 in. outside diameter. The inset is an enlarged view of a piece cut from the ring, to show the cross section, and particularly the rim, or "spoiler" at the outer edge.

made in that way holds quite well for the whole range of speed! The great increase in flying distance over that of the Frisbee attests to the smaller amount of turbulence as well as better aerodynamic design. But the challenge of solving the problem without *any* turbulence is still there, and the *Guinness Book* awaits the result.

The Aerobie is quite an achievement—but where do people play catch with an object that flies over 300 m? Scott Zimmerman used the Rose Bowl for his record throw.

References
1. *Phys. Teach.*, **21**, 325 (1983).
2. Registered copyright. Sold by Superflight, Inc., 81 Encina Ave., Palo Alto, CA 94301. The company was helpful in supplying information and samples.
3. Stanford University, Stanford, CA 94305
4. The action of spoilers can be seen any time an airliner lands. The instant the wheels touch the runway, spoilers on the top surfaces of the wing are raised to kill the lift and put the weight on the wheels.

How Physics Is Used to Repel Deer and Fleas

In Michigan, as in other places where deer are numerous, the animals crossing the highway at night are a real hazard to travelers, causing expensive damage to the vehicles and injuries, sometimes fatal, to people. It has been found, by tests, that deer can hear and are alarmed by sound frequencies just above those that can be heard by (and are annoying to) adult humans,[1] namely the range 16 to 20 kHz. That fact is being made use of in greatly reducing the deer-related accidents on the highways.

The device that does the job is simple. In essence it is a funnel, about 2.5 cm (1 in.) in largest diameter, mounted on the outside of the vehicle (one on each side), with the large end facing forward into the wind. The constricted end is shaped that it whistles at the required high frequency. The whistling occurs at speeds over 30 mph. The whistles are marketed in the U.S. under the name Save-A-Life.[2] They come in an enclosure that looks like a miniature jet engine, chrome-plated, at about $30 per pair (Fig. 1).

So far the application of the whistles has been mainly to fleets of freight-hauling "semis," highway patrol cars, and other vehicles that travel at high speed at night. As an example of successful use, Meijer's Inc. ("Thrifty Acres") has equipped its approximately 100 semitractors with the whistles. The fleet's 30–35 deer-related accidents per year has dropped by 80%. Dollarwise, the figures are no less impressive: The damage to a semirig can run to $20,000.

The business of repelling not only deer, but all manner of creatures, has taken a leap with the advent of compact, microchip equipment for generating high frequency power which can be turned into sound by miniature loudspeakers. One sees ads for above-audible sound generators for chasing away mice, cockroaches, stray dogs and cats, and flying insects. A general rule,

Fig. 1. The whistles in their pods, one for each side of the vehicle (Photo supplied by Save-A-Life, Inc.)

that has a basis in physics, is that the smaller the creature the higher will be the frequency it can sense (by one means or another). If the frequency is high enough, even fleas are disturbed. Witness the latest entry, featured in a full page in the *Sharper Image* catalog,[3] an intermittent ultrasonic generator built into a dog or cat collar. It makes fleas leave the animal. One might be concerned as to where the fleas would take refuge—the catalog does not say!

References
1. The qualification "adult" is the author's. The range of hearing of small children can reach 16 kHz. But, if, as probably is true, the device to be described is use mainly on trucks, police cars, etc., the matter may not be important. The advertising literature does not raise the question.
2. Save-A-Life, Inc., P.O. Box 1226, New York, NY 10025.
3. Issue of Aug., 1986. 680 Davis St., San Francisco, CA 94111.

How Lumps of Steel Shape Themselves into Near-Perfect Spheres

The little steel balls that are used by the billions in the bearings of all kinds of machinery and appliances are marvels of precision. ''Garden variety'' balls are precise as to diameter and sphericity to 0.00006 cm. Premium grade balls are up to ten times as precise. That such precision can go along with manufacture in huge quantities at low cost is most surprising. It is true only because of one special property of a sphere: Under quite simple conditions of lapping,[1] a rough lump of material will wear away into an increasingly near-perfect sphere.

There is evidence that the ability of rough lumps to wear themselves into spheres was discovered and put into commercial practice about three centuries ago.[2] The purpose then was not to make bearings but to turn pebbles of colorful stone into polished spheres for use as toys or decorative items. The basic method works as well today as it did then, and it is at the heart of modern manufacture. In recent times only one new feature has been added to the process—a feature that makes possible the production of very large batches of balls, all having precisely the same diameter. It will be best to start by describing the basic part of the process and then the mass-production feature. The idea can be gained from the small model in Fig. 1, constructed from information in the article of Ref. 2.

In modern manufacturing[3] the lapping is done by a pair of circular steel plates, called rill plates, as drawn[4] in Fig. 2A. On their opposing faces are shallow, concentric grooves (hence the term rill). At the start of a run, the grooves (rills) are filled with steel balls that have been pre-prepared roughly spherical and heat treated, by manufacturing methods to which we will not devote space here. The plates are forced toward one another, while one of them rotates. The balls roll in the rills, each completing a trip around the fixed plate while the moving plate makes about two revolutions. Abrasive powder suspended in oil or water is fed in. Lapping takes place because, although the main movement is rolling, parts of the surfaces of the balls are always sliding. Of course, as the process goes on, the rills wear deeper, so the plates have to be re-faced at intervals. That is why the rill plates are separate and removable from the spindles, as indicated in the figure.

So why does that make the balls into near-perfect spheres? The crucial point is that in the process each ball continually changes its orientation, in a *nonrepeating* way. That is true because the groove is curved: The curvature gives the ball some turning, or rotation, around axes other than the main axis of rolling. Those

Fig. 1. A model, in miniature, of the type of ball mill used as early as the late 1600s in Austria. One view (top) shows the concentric circular grooves in the thick disks, with two of the grooves in the lower one filled. In the other view (bottom), the upper disk is in place for running. Water from the sloping trough hits the fins on the top of the upper disk, to make it turn. In the model shown, both disks represent slices of a tree. A block of sandstone was often used in the lower position. In either case, the grinding was done by sand and water. The pebbles chosen to be ground into spheres were of the softer natural materials, e.g., marble.

rotations and the rolling rotation are not commensurable, so no orientation repeats exactly. The only shape of object that can fit between the plates, in the grooves, *in all orientations* is a sphere. And, that is exactly the shape into which the grinding makes it! For example, if a ball is a little egg shaped, the material on the high sides will get a little more grinding pressure and be reduced. And if a ball is larger in diameter than the others in the same groove, it will receive more grinding pressure and get smaller faster.

The process so far described will produce balls of uniform size in each of the circular grooves. But because grooves wear to different depths, the balls can end up in as many sizes as there are grooves. There might be a 100 balls of each size—but that would be of no use in modern mass production. The feature added in modern times to solve that problem is the circulation of the balls to and from a large holding bin, or hopper. That accomplishes two things: It shuffles the balls among the grooves, to eliminate the problem of different wear, and it allows production runs to be of unlimited quantity.

Here is how the circulation works. There is an open slot in the rill plate that does not rotate (Fig. 2B). A scoop(s) removes the balls from all the grooves as they arrive at the slot (Fig. 2C). They are conveyed to the hopper and put in at the top. Balls from the bottom of the hopper are conveyed back and inserted into the grooves. The process is continuous. To trace a given ball:

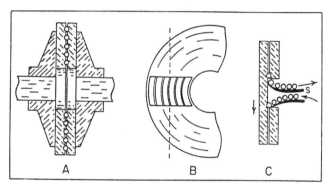

Fig. 2. A. A pair of modern rill plates, with balls in the grooves between them. The plates can be removed from the spindles, so a cut can be taken off their surfaces when the grooves become too deep. B. View of the nonrotating rill plate showing the open slot, or window. Through the window, the rills in the rotating plate can be seen. C. A cross-section view, at the position indicated by the dashed line in B. It shows the scoop, removing the balls as they arrive at the window, to be conveyed to the hopper and the guide that puts the balls back into the grooves as they come from the hopper. The rill plate on the left is the rotating one.

After making one trip around a groove, it goes to the hopper, where it resides for an average time depending on the content of the hopper. When it returns, which groove it gets into will be a matter of chance. So after sufficient running time, *all* of the balls will be of the same diameter.

A few numbers will be of interest. Typical disks are from 1/2 to 1 m in diameter. The number on concentric grooves depends on the diameter of balls to be lapped;

the whole area is used. The disk that rotates goes at 1 to 2 rev/s. The hopper holds about 500 kg of balls. If they are 1 cm in diameter, there are 100,000 of them. The number being lapped at any instant can be estimated. If the disks are 1 m in diameter and the area is 1/3 covered by balls, there are 2000. Thus, only 2.5% are being lapped at a given time. That is consistent with the fact that the running time for a batch is something like 8 to 12 hours, which would give the order of 20 min lapping for each ball.

To bring the balls to their final precision in diameter and surface finish, they are often put through another run in rill plates with finer abrasive. And for a high polish, they may be run in rill plates with fluid but no abrasive at all. To achieve the precise size, the diameter of the balls is monitored, first with calipers, and in the final stages by an optical method. For the premium-grade balls, temperature expansion becomes important—samples have to be removed to a controlled-temperature chamber for measurement. One person I talked with remarked that ball making is one instance in which the precision of the process exceeds the precision with which the product can be measured!

Ball bearings with which we come in contact with—found in sewing machines to automobiles—contain balls generally in the range from a millimeter to a couple of centimeters.[5] But for heavy machinery, balls up to 10 cm are stock items. Much smaller and much larger ones are made. The smallest ones can best be seen with a magnifying glass. On the large side, some years ago a 30-cm steel ball was donated to the Washtenaw Community College by Industrial Tectonics Inc., for a Foucault pendulum. (Later it was stolen!) A local story has it that the balls used in the bearings of the "Big Bertha" gun that bombarded Paris from 120 km in World War I were made in Ann Arbor and obtained by the enemy by a circuitous route.

References
1. The term lapping, rather than grinding, is generally used when the abrasive is in the form of powder or granules, suspended in oil or water. It is introduced between matching surfaces of two pieces of material, e.g., glass or steel, as they are moved with respect to one another.
2. "Balls of Marble and Steel" in *Styr-Actuell*, No. 2, 1964. (No author indicated.) Reprint supplied by Industrial Tectonics Inc. The earliest date, 1680, was found in tax records.
3. For facts about the manufacture of balls, I found no lack of local experts. Ann Arbor has a long history as a source of ball bearings and of balls not only of steel but of other materials for special purposes. Helmut Stern of the Arcanum Corp., Don Allum of NSK, Inc., and Alden Cook (retired from) Industrial Tectonics Inc. were especially helpful.
4. This follows illustrations in "Fundamentals of Ball Grinding," by Robert N. Kopp, *Abrasive Engineering*, pp. 22–24, Jan./Feb. 1973. It is known as the Hoffman process.
5. In manufacture and trade, balls for ball bearings go by exact integral inches and fractions of an inch.

The Transponder: Device of Many Uses

The Latest—Automatic Focusing of the Movie Camera

Remember the discussion of the popular automatically focusing personal 35-mm cameras?[1] Automatic focusing of a more sophisticated kind has entered the movie-making industry, and it accounts for some shots that would be next to impossible without it. An example would be the problem of following an actor running and dodging through a crowd and keeping him in perfect focus all the time. In the new scheme, a microwave radio transmitter at the camera sends out continuously a series of pulses—short trains of waves. The runner has on his person a match-box size *transponder*, a device that receives the pulses and retransmits them instantly, shifted in wavelength. The equipment at the camera receives the retransmitted pulses, and in its microprocessor circuits translates the time difference between the outgoing and returning pulses into a distance setting for the lens. Everything is automated right down to the motorized focusing movement of the lens—everything, that is, except for pointing the camera. (That, at least, remains as something for the cameraman to do!) Not only does the focus follow almost instantaneously, it is single-minded—ignoring everything but the actor who has the transponder.

My information comes from Robert Bogle, long ago my Ph.D. student, later my colleague in research, and in recent years a high-tech entrepreneur. He and associates have pioneered an autofocus system for the movie industry called RADARFOCUS.[2]

A radar-like system for movie work would not be possible without the recent development of extremely fast-acting solid state components. The velocity of light (and of radio waves) being 3×10^8 m/s, we note that the time separation between a transmitted pulse and the return, either by simple reflection or by transponder, from an object at 150 m distance, is 1 μs. Not long ago, ranging at a distance that short was "high tech" in the radar art. But it would not have been short enough to be of use in moviemaking. Things have changed. Bob Bogle tells me that with the fast solid-state components now available, the minimum ranging distance has come down to about 1 m. At that distance, the time separation between transmitted pulse and return is about 7 ns (nanosecond, 10^{-9} s).

The short ranging distance places some severe conditions on the radio frequency, or wavelength. If the time at which the transmitted pulse starts toward the transponder and the time the return pulse starts to be received are to be compared to an accuracy of let's say 10 ns, then the turn-on times, or "rise times" of those signals have to be that sharp. It takes a number of cycles, maybe 100, for a transmitter to come up from nothing to full power, and it takes a similar number of cycles for a receiver to recognize the return signal. Thinking in terms of orders of magnitude, we find that to have 100 cycles in 10 ns requires that the frequency be 10^{10} Hz (10 GHz). And that is, in fact, the order of frequency that is used. In Bob Bogle's system the transmitted frequency is 10.2 GHz, and the return from the transponder is 20.4 GHz.

The radiation is transmitted and received by a horn, as is common when the wavelength is very short (Fig. 1). From the photograph, we can make an estimate as to how well the radiation is directed to the subject. The wavelength at 10.2 GHz is 3 cm. The horn appears to have a maximum width about 3 times that. Thinking back to the simple diffraction laws, we find that the ranging system should "see" in a cone of about 20 deg (full width)—again, satisfactory. However, the sharpness of pattern is not critical, since the system ranges only on the actor carrying the transponder. Sharpness does conserve power and increases the signal strength.

When you see your next movie, take note of the focus during fast-action shots.

Transponders have been around for a long time and have many applications. Some additional examples will be of interest. In World War II, combat airplanes were equipped with transponders. They were tuned to be triggered by (friendly) radio or radar signals. By the character and/or shifted frequency of their return signals, they identified the airplane so it would not be shot down by fighters of its own side. The system came to be called IFF (Izzy Friend or Foe).

The most interesting application of the transponder that I know of is in the synchronization of clocks over long distances. I first learned something about this application while visiting the National Bureau of Standards at Boulder, Colorado. One of the functions of the NBS is to keep highly precise time and to communicate it to far-away stations. It is done by radio.

If you have an all-band receiver, you have heard the time signals which are sent continuously from WWV on multiples of 5 MHz. The format is the same as that used on the telephone: "At the tone the time will be . . . beep".

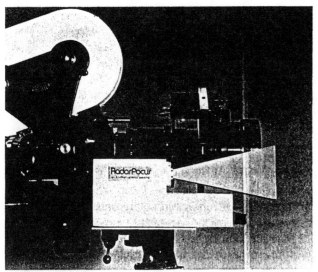

Fig. 1. The microwave transmitting and receiving horn, mounted on a movie camera. (Photo courtesy of R.W. Bogle.)

A problem arises when ultra high precision is needed, as for synchronizing atomic clocks, separated by great distance. It is then necessary to correct for the time of travel of the radio beep, which in this case is a very sharp pulse. There is a clever way of finding the exact time of travel, without having to know the distance between stations or what influence the earth, air, or ionosphere may have on the travel time. The idea in its simplest form is to send the clock time and a beep from station A to station B. A transponder at B returns the beep to A. The round-trip travel time for the beep, measured at A, divided by 2, is the correction to be applied to the transferred clock time.

The modern method[3] employs a transponder carried on a geostationary satellite.[4] The time pulses (microwave, 20–30 gHz) are sent between ground stations A and B via the transponder, both ways. In practice the routine of the transmission is, as you would expect, a little more complicated than the sketched in the paragraph above, but the basic indegredient is still the measurement of the two-way travel time of a pulse. The precision in transferring time, almost irrespective of distance, is of the order of 10 ns. To be impressed, note that light goes only 3 m in that time!

References
1. *Phys. Teach.*, **22**, 600 (1984).
2. Worth Associates, 991 Skylark Drive, LaJolla, CA 92037.
3. Information was supplied by Karl Kessler of the NBS, Gaithersburg, MD 20899 and by Wayne Hanson of the NBS, Boulder, CO 80303.
4. A geostationary satellite is at such radius that its orbital period is the same as the period of rotation of the earth. That makes it fixed in position as seen from the earth. There are many such satellites· in the sky, used for television, telephone, and other purposes. Many carry transponders. Calculation of the orbit radius is left as an exercise!

Three-Dimensional Moiré Patterns

Recently, I wrote about a wing-flapping plywood goose that has served me as a conversation-generating home ornament.[1,2] The interest it produced encourages me to describe a device of similar purpose—one from which I have had much "mileage" in terms of trying to explain the physics to nonphysicist visitors. It starts with the familiar moiré effect[3] and goes a step further, with the object of causing maximum confusion to the observer's perception of depth—stereo vision. An exhibit I saw some years ago puzzled me enough to start me experimenting with the effect, the end results of which may or may not be very close to what I saw. But better, I think!

The standard way of observing the moiré effect is by superposing two families of lines, circles, grids, or other figures. The one on top is on transparent plastic; the one underneath may be on paper or plastic. If one slowly slides the top one around, artistic, changing effects are produced. Moiré effects from *three-dimensional* configurations are not uncommon. If the first, transparent, sheet is moved out to 10 cm or so in front of the second sheet, it can remain fixed, and the pattern will change when the head is moved—a matter of parallax.

So far, so good. The device to be described follows the three-dimensional configuration, but attention is paid to a set of a parameters normally ignored, namely the separation between the observer's two eyes, and the distance from the eyes to the exhibit. With those quantities, the geometry of the device can be worked out to produce *maximum depth confusion* (which of course equals maximum attention-getting).

Fig. 1A gives the idea of the arrangement, by showing the lines of sight of the two eyes (a) of an observer. A black card (b) has in it regularly spaced vertical slits. It is at some distance in front of a set of vertical white lines (c) which are on or in front of a black background. (The white lines are indicated by dots.) We pick a distance of, say, 2 m from the observer to the exhibit, and then satisfy two conditions. The first is that the lines of sight (shown as solid lines) from the two eyes going through neighboring slits should end at the same white line. That determines the spacing of the slits. The second is that the lines of sight shown as dashed lines going through a different pair of slits should also end on a white line. That determines the spacing of the white lines.

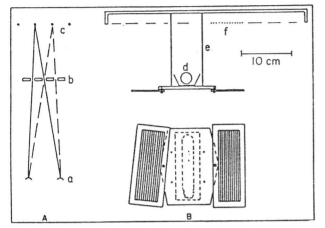

Fig. 1A. The geometry of the viewer's eyes (a), the slits (b), and the white lines (c). It is not drawn to scale.

Fig. 1B. Top and front views of the assembled device, to scale.

With the above conditions satisfied, a lateral movement of the head will make the whole area appear alternately black and white. But the best effects occur if the observer is at a little more or less distance than that which satisfies the above conditions, and/or the slits and the lines are not quite parallel. Then there are broad light and dark bands that move sideways or up or down as the head moves, and are not seen in the same places by the two eyes. That is where the confusion begins for the sense of depth.

At this point let's explain the physical arrangement, especially the illumination, which is essential for a good effect. Fig. 1B is a scale drawing of an arrangement that works well. The slit and line spacings are calculated for an observer distance of about 2.5 m. Any arrangement that suits the artistic whim will work, if only the two simple rules for the spacings are observed. Illumination is important. The lamp (d) is the small tubular, straight filament 25-W type, used for lighting pictures, music stands, etc. It is shielded from view from the side by a metal trough. Each frame of slits is mounted with a single screw, so that it can be set at any small angle to the vertical, to give interesting effects. The light and slit assembly is held by four stiff wires (e in the figure); f

is a sample of an array of strings that serve as the white lines. The whole device is hung on a wall, like a picture, preferably in a location of dim light (Fig. 2).

For the serious builder, some comments about making the slits and white lines may be useful. For the white lines, a shallow plywood tray, 40 cm square and 2.5 cm deep, was made and painted flat black. Uniformly spaced notches were made in the rims of opposite sides, and white nylon string was laced in the notches. The diameter of the string should be about half the open distance between strings. At first the arrays of slits were made in much the same way: by lacing frames with black yarn. Later, slits were made in 1.6 mm (1/16") model builders' plywood, with a thin circular saw. The width of the slits should be about half the diameter of the yarn or of the solid plywood between them. A way not tried is to cut and glue narrow strips of black paper to the frame. The spacings of the white lines and slits are purposely not specified; they should be found using Fig. 1A and the builder's own viewing distance, etc. Note that there is no space for a socket for the lamp bulb. It is held in a clip and the wires are soldered to the base.

Have fun.

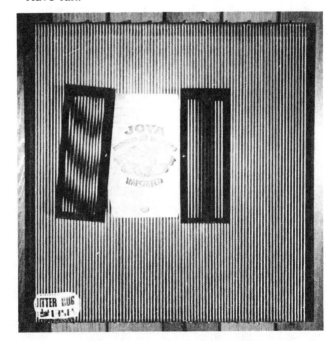

Fig. 2. The device mounted on the wall. To take the photograph, extra light was projected onto the front. Without that, the moiré patterns would have been relatively brighter. The distance to the camera was a little less than that which would conform to the condition of Fig. 1A. If the camera were moved the distance between an observer's eyes, the vertical band would be displaced laterally, and the bands at the left, up or down. (You guessed it: the piece in the center is a cigar box lid.)

References

1. *Phys. Teach.* **24**, 169 (1986).

2. The goose has offspring: The Edmund Scientific Co. (101 E. Gloucester Pike, Barrington, NJ 08007) 1986 annual catalog lists variations that work in the same way: a seagull and a flying machine from the age of dinosaurs, the pterodactyl. If you want to build one on prehistoric lines see Smithsonian for Mar. 1986, pp. 72–84.

3. Moiré is the term for a weave and finish of cloth that makes it seem to have waves or ripples—used mostly for silk (*Encl. Britannica*).

Why Your Gas Tank Doesn't Overflow

How does a gasoline pump know when your tank is full so it can prevent overflow? That question was asked by Steve Walker, a student at Acalanes High School (Lafayette, CA 94549) and communicated by his teacher, Raleigh Ellison. The same was asked by Stuart Leinoff of Adirondack Community College (Glen Falls, NY 12801). With help from Michael Rayne of the Wayne Division of Dresser Industries Inc. (124 W. College Ave., Salisbury, MD 21801), and a little surreptitious experimentation at a service station, I can give the main ideas.

First of all, there is nothing in the main enclosure that can sense when your tank is full. Its machinery pumps gasoline from the storage tank as required to maintain a constant positive pressure of gasoline in the hose all times. The mechanism that starts and stops the flow, and senses when the tank is full, is entirely within the pistol-shaped head at the end of the hose, the nozzle of which you push into the filler pipe of your tank. Of course other useful functions are performed by apparatus inside the main enclosure, such as metering the gasoline pumped and telling you how much you have to pay. For those service stations that use a prepay system, electric power is not supplied to the pump that keeps the hose under pressure until the money is in hand.

The next time you get gas, look on the outside of the nozzle a centimer or two from the end, and you will see a hole a few millimeters in diameter. A small tube runs from that hole, inside the nozzle and hose, to a suction pump in the main enclosure. Inside the "pistol," a branch of the tube connects to a small cup covered by

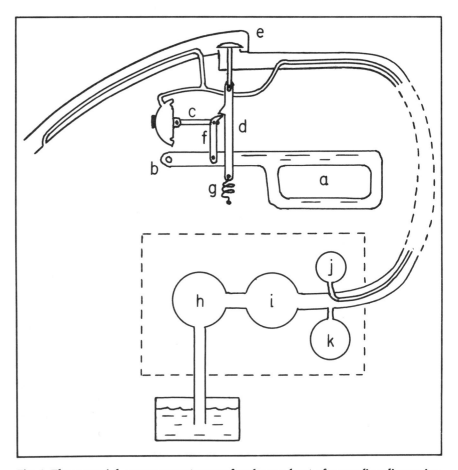

Fig. 1. The essential components (except for the readout) of a gasoline-dispensing system. Above the dashed rectangle, the parts of the "pistol" at the end of the hose. Inside the dashed rectangle, the components that are in the main enclosure: gasoline pump, meter, vacuum pump, and pressure switch. At bottom, the underground tank.

a flexible diaphragm (Fig. 1). When the tank is full, and gasoline instead of air enters the hole, the pressure in the tube goes more negative with respect to atmospheric. (That is, because gasoline offers more resistance to flow than does air.) The lowering of the pressure causes the diaphragm to pull in, and that furnishes the mechanical force necessary to trip a latch that is holding the gasoline valve open. So the flow stops, instantly.

If you want to verify the above, you may have to wait until you fill a can with gas for your lawn mower or snow blower. You can't do it while putting gas in your car,

---

if it is a modern one. There is a safety vane in the filler pipe that must be pushed open with the nozzle to allow the gasoline to go into the tank. That puts the end of the nozzle out of reach. While filling the can for my mower, I repeatedly put my finger over the small hole in the side of the nozzle. Each time, the latch tripped, closing the valve and stopping the flow, just as if the tank were full.

A system with the bare essential working parts is diagrammed in Fig. 1. (While the functions indicated are the true ones, the forms and arrangements of the parts are drawn only with simplicity in mind.) In the upper half of the figure are the parts in the "pistol" at the end of the hose. Inside the dashed rectangle are the parts in the main aboveground enclosure. Below that, the underground storage tank is indicated. The dollars and gallons readouts are not indicated but will be mentioned later. When the handle, **a**, is squeezed upward it pivots about **b**, raising the part **f**, which engages the tooth on part **d** and pushes it upward. That pushes the valve **e** up, letting the gasoline flow. When gasoline enters the small tube at the end of the nozzle, the resulting "vacuum" causes the diaphragm to pull in, and through the link **c** pull **f** to the left, unlatching **d**, letting the spring **g** pull the valve closed.

In the main enclosure there are four principal parts of the fluid system: a pump, **h**, a meter, **i**, a vacuum pump, **j** and a pressure-sensitive electrical switch, **k**. When the customer starts the flow out of the nozzle, the pressure in the hose starts to fall, causing the pressure switch to close and start the pump that keeps the hose

full. It also starts the vacuum pump. The meter is really a piston-cylinder "engine" that is run by the gasoline that flows through it. That type of meter is used because it passes a precise volume of fluid each revolution. It has to have three cylinders, so it will always start—not be caught on dead center. If the gallons/dollars readout is mechanical, the "engine" simply drives a mechanism having number-cylinders that read the dollars, etc. If the readout is electronic, as is the trend, the engine opens and closes a series of contacts on each revolution, sending electrical pulses to whatever computer system there is, which, in turn, counts them and computes and displays the cost etc.

A final point—something that seemed puzzling—about the vacuum pump. I tried a little experiment, doing things in reverse: I held my finger over the hole and *then* squeezed the handle. Gasoline started to come out, but the latch tripped and stopped it before much more than 100 cc had come—less than half a second it seemed. How can it be so fast if the vacuum pump only starts to pump out the tubes and the diaphragm cup after the hose-pressure switch closes? So another test: Feeling around the hole when the handle was not being squeezed detected no suction. Evidently the pump was not running. Back to Michael Rayne. He confirmed that the vacuum pump is started by the hose-pressure switch. Still surprising.

You may be tempted to experiment. If so, Michael Rayne warns, don't spill gasoline around the pumps and cars: It's more dangerous than you think.

The Hydraulic Ram: How to Make Water Go Uphill

In my early days, a house I lived in was supplied with water from an elevated tank that was kept full by a hydraulic ram. Later, after learning some physics, I gained much admiration for the ram as a device for raising water with almost no moving parts. But alas, like that other great simple invention for raising water, the windmill, the ram now is seldom seen. My purpose in looking back at the ram is to show its novel use of a few simple physics ideas—certainly not to prepare you to install one for supplying water to your house. In particular, the acceleration of water in an inclined pipe and the elastic collision of columns of water are demonstrated.

To get started, we may as well go back to the first proposal of the idea, which was made in 1772 by an English watchmaker named Whitehurst.[1] He

demonstrated that if a source of water at a small height above the ground was available, a fraction of it could be pushed up to a much greater height, using nothing more than valves. In Fig. 1, **A** shows the system and **B** shows the ram in detail. A long, downward-sloping pipe comes from the source of water—a creek or spring—and ends in a gate, shown here as a stopcock. An alternate exit for the water is through a one-way *check valve* (a) into a closed chamber (b), called an *air bell*. A pipe from the chamber runs up to an elevated tank.

Let's see how the system can put water into the tank. We start (left detail) with the stopcock open. The vertical pipe is full of water. The air and water in the bell are under a pressure determined by the height of the vertical pipe. The pressure keeps the check valve closed. The water in the sloping pipe is running out of the stopcock and gaining velocity. We suddenly close the stopcock (right detail). The water takes the alternate exit by pushing the check valve open, flowing into the bell until its kinetic energy is spent in further compressing the air. When the flow comes to a stop, the check valve closes. The air then expands back to its former pressure by pushing out, into the vertical pipe, the amount of water that came in. Thus water is put into the tank.

With important qualifications to be brought out later, the foregoing process can be classed as the elastic type: energy conserving (like the collision of two carts on a track, with a steel spring between them to store the energy temporarily). Only here, instead of the spring, there is the compressible air in the bell. The energy gained by the water moving down in the inclined pipe should be equated to that used in raising water from the ram up to the tank. If the fall in the sloping pipe and the height up to the tank are in the ratio of 1 to 10, then 1/10 of the water that leaves the source should arrive in the tank.

Whitehurst filled his tank the hard way—by manually operating the stopcock. But before long it was automated and by none other than Joseph Montgolfier who, with his brother Etienne, had invented the hot-air balloon and staged the first manned flight in one.[2] To the ram Montgolfier added two features: an automatic valve (called a *waste valve*) to replace the stopcock, and a small check valve (c) called a *snifter* to maintain the proper amount of air in the bell. In those strokes he transformed the ram from a curiosity to a much-used device that could keep a tank filled for years without attention.

The details of Montgolfier's ram are shown in Fig. 1C. The waste valve is a disk on a spindle, which by gravity rests in the position shown in the left detail, until the velocity of the water flowing around it is great enough to lift it. It then closes (with a bang) and water flows through the check valve (right detail), until its energy is spent. What happens in the next fraction of a second made both of Montgolfier's improvements possible. The moment the water stops flowing forward through the check valve, there begins a rebound: The compressed air pushes water backward through the check valve, during the short time it takes the check valve to shut. Water in the sloping pipe is of course also pushed backwards. When the check valve reaches the closed position the backward motion has to stop, and that produces a momentary reduction in pressure in the water below it and in the waste valve. Actually the pressure dips below that of the outside atmosphere. In that instant of reduced pressure the waste valve is able to start dropping down, and a little air is able to come in through the snifter check valve, to be carried up into the bell on the next cycle.

A word as to why the snifter is so important. The compressed air in the bell dissolves, slowly, in the water with which it is in contact. Without replenishment by the snifter, there soon would be no more air in the bell. If the snifter admits air in excess of the need, bubbles will go up the vertical pipe. Thus, it is self-adjusting.

Parameters for the ram that are typical of those actually used will be of interest. They might be: a vertical rise from the ram to the storage tank of 15 m, a length of the sloping pipe of 20 m, a diameter of 30 mm, and a fall of 2 m. The time of a cycle is short: It might be 4 s, the water accelerating in the sloping pipe for 3 s, and the waste valve reopening in the fourth second.

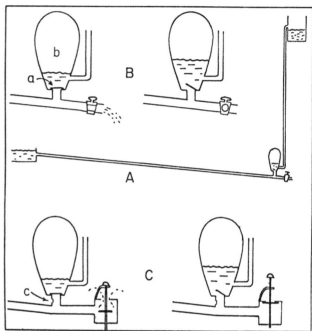

Fig. 1.A. The overall layout: the source at the left (spring or creek), the ram in the center, and the elevated storage tank at the right. Fig. 1.B. The ram as used by Whitehurst in 1772. It has only three parts: a check valve (a), an air bell (b), and a stopcock. As shown at left, the water is accelerating in the sloping pipe. At right, the stopcock has been closed, and the water is entering the bell through the check valve. Fig. 1.C. The ram as improved by Montgolfier. At left the waste valve is open, and water is accelerating. At right the waste valve is closed, and water is pushing through the check valve into the bell.

With the above numbers we can make a calculation of the performance. We will have to do so on the premise that the system is energy efficient, while knowing of several reasons why the result may be rather far from the truth. Mainly we have to leave out of consideration the energy loss due to fluid friction, the kinetic energy that is lost in the water that spews out of the waste valve during the acceleration, and the fact that the water in the sloping pipe does not move as a rigid body. (The water in the center moves faster than that at the walls.) But let's do it anyway, and see what we get.

Just equating energy, 2/15 of the water flowing down the sloping pipe will be raised to the tank. But how much comes down? In each cycle the water starts from rest and accelerates at 1/10 g for 3 s. That gives it an average velocity of 1.47 m/s, and in the pipe of 30 mm diam that is a flow of 1.04 l/s. So, in 3 sec, 3.12 l comes down; 2/15 of that, or 0.416 l, are put into the tank. Multiplying by the number of 4-s cycles in 24 hours, the tank gets nearly 9000 l. Enough for quite a few showers!

In an old hydraulics handbook, a formula is given for the practical efficiency of the ram.[3] On breaking it down, the formula amounts simply to saying that 2/3 as much water is raised to the tank as would be raised if the energy efficiency were 100%, as we assumed in our calculation above. Tables are also given, for tanks at different heights, etc. All the figures suggest efficiencies of around 50%. Whether the numbers are measured or calculated is not told; neither is the number of seconds per cycle. We assume that the figures are given for the optimum cycle. The only experimental evidence found (dubious at that) was a statement by a Mr. Millington, in England, that his ram raised "134 hogsheads (6300 gal) to a tank 134 ft high from a source 4.5 ft high in 24 hours."

A comment about the time for a cycle, the information that was missing from the tables. The simple kind of calculation we made (losses omitted) would indicate that if the cycle were lengthened, more water would be raised to the tank in a day. That's because the average velocity of the water in the sloping pipe would be higher. But the losses, especially fluid friction, increase rapidly with the velocity. In fact, if the waste valve were left open, the velocity would quickly reach a steady value, such that all the energy gained in the fall would be going into the several kinds of loss. So, clearly, there is an optimum cycle time. Experience has shown that it is a very few seconds. It is made optimum by adjusting the weight of the moving part of the waste valve and the position to which the disk drops and sits when the valve is open.

References
1. Ewbank's Hydraulics and Mechanics (Bangs, Platt & Co., New York, 1850).
2. Ewbank's Hydraulics and Mechanics.
3. Andel's Pumps, Hydraulics and Air Compressors. Edited by Graham. No further information—pages missing.

Penguins on an Escalator

Why is there such fascination with penguins? As soon as I laid eyes on this toy, operating, in the San Jose, California airport, I started reaching for my wallet. As I watched it work, I was stumped by two questions of physics. How were the penguins made to (seem to) climb stairs, and how were they made to run smoothly and fast down a track having only a gentle slope? Clearly, the only solution was to bring the gadget home and do "reverse engineering" on it.

The toy is shown in Fig. 1.[1] For scale, the highest point is 23 cm above the tabletop, and the penguins are 4 cm tall. The track is 160 cm long, with an average slope of 5.6°. The material is plastic. The penguins go up the 11 steps in 7 s and make the trip down the track in 3.5 s. Rather loud squawks are emitted from somewhere within, at 4 per s. The motor that makes everything go is powered by one D-cell. The current drain is 300–400 mA, roughly the same as for an ordinary D-cell flashlight.

Let's first inquire as to how the penguins get from the bottom to the top. The arrangement has the appearance of the well-known people-escalator. There is a set of stair steps and a rail on each side of the rider, but there the similarity ends (Fig. 2). The stair steps do not move. The side rails are not straight but sawtooth in shape. The penguins do not stand on the steps but are supported

on pins that stick out from the sides of the body. (The pins show clearly in Fig. 3.) The parts that lift the penguins are a pair of clear, flat plastic pieces, one on each side between the handrail and the penguin's body (best seen in Fig. 2, second from the top). Their upper edges are of the same sawtooth shape as the handrails. By merely *oscillating*, they make the penguins progress

Fig. 1. The "Playful Penguin" toy. The top is 23 cm above the table, and the penguins are 4-cm tall. The main material is plastic. After being raised to the top with squawking sound effects, the penguins go down the 1.6-m-long track to start over. The interesting physics is in the method by which they are raised.

Fig. 2. The mechanism by which the penguin is raised has been stopped at successive stages, to show a complete cycle.

Fig. 3. Two features are shown: the side pins, by which the penguins are supported while being raised, and the solid brass rollers by which they travel on the track. Note the very small diameter pins on which the rollers turn, giving low friction.

upward. How that works cannot be seen when the toy is running; the motion is too fast. To figure it out, one has to be able to stop the motion at various stages—something not easy to do in an airport gift shop! And that is one of the reasons I brought it home!

Fig. 2 shows the motion stopped at four successive stages of raising the penguin from one sawtooth (of the handrail) to the next. (Start from the bottom picture.) The oscillating motion of the clear plastic sawtooth is *parallel* to its (and also the handrail's) upward sloping edge. Also, the position of the set of clear sawteeth is such that its teeth are midway between the teeth of the handrail. That is seen best in the second picture from the top, Fig. 2.

Now let's follow the motion, starting from the bottom picture in Fig. 2. The penguin's side pins are in the crotch of the siderail. (The clear sawtooth that is visible in this picture is the one on the far side of the penguin.) Next picture, the clear sawteeth (both sides of the penguin) have risen, and because of their half-tooth displacement, they have lifted the penguin by its side pins. The clear sawteeth continue to move upward, and in the next picture (third from the bottom) the down-sloping edges of the clear sawteeth are *above* the down-sloping edges of the handrail. That has allowed the penguin to slide, by gravity, to where its side pins rest in the crotches of the clear sawteeth. Now the clear sawteeth reverse and go down. When they have gone down *more* than a full tooth length, the side pins are free to slide the rest of the way down to the crotch of the siderail. But that crotch is one higher than the crotch they occupied in the bottom picture. Then the clear sawteeth start up, for the next cycle.

Fig. 3 shows, besides the side pins on the penguins, the rollers. The latter are solid brass, heavy enough to place the center of gravity below the side pins. The track fits the rollers exactly, so the penguin stays pointed in the direction of travel. Friction is very low: In a test, the penguins rolled at constant speed down an incline of only 2.8°.

Earlier it was mentioned that there are stair steps, on which the penguins do not stand and which therefore seem, at first sight, to serve no purpose. But, they do have a function. In Fig. 2 it will be noticed that the penguins bob back and forth from the vertical, pivoting on the side pins, as they "climb." The weight distribution and location of the pins is such that the penguin, supported only by the pins, leans forward. As the penguin moves along, each stair step gives it a little push at the lower end, making it bob forward.

The squawks are produced by a small bellows (3 cm square) that blows through a reed, into a horn about 3 cm long. The squeezing of the bellows (4/s) is, of course, done by a link to the motor and gear system.

One cannot help being impressed with the design of the toy. Many of the features are clever, and the construction is unusually sound. For example, the two halves of the main enclosure are put together with 6 Phillips screws (For the convenience of inquiring physicists!). Most toys are simply glued.

I end by suggesting a problem: The penguins take 3.5 s to go down the inclined track (length and average slope already given.) How does that compare with the ideal case (uniform slope, no friction) for a rolling solid cylinder and for a sliding block? The penguin is somewhere between: Most of the mass is in the rolling cylinders. It's analogous to the textbook cart with massive wheels, on an incline

Reference
1. "Playful Penguin Race" by Dah Yang Toy Industrial Co., Ltd. (Taiwan), $20.

Body Volume Measured by Sound Waves

No doubt you have made sound by blowing across the mouth of a jug or a soft drink bottle and have noticed that if the bottle is partly full of liquid, the tone is higher than it is if the bottle is empty. With proper calibration, the volume of the content can be determined from the frequency. W. Gregory Deskins and colleagues at Hoover Keith and Bruce Inc. are developing essentially this same method for measuring volume of the living body.[1] I am much indebted to Greg for descriptive material[2,3] and discussion about the project.

Body volume, from which average density can be found, is of much interest for some kinds of medical diagnoses. Developing a method for finding the volume by means of sound in air would allow researchers to avoid some of the serious drawbacks of the "textbook methods" which typically involve immersion of the subject in water. As examples of those methods: 1) the subject is lowered into a full container of water, and the amount caused to overflow is measured; or, 2) the subject is weighed when submerged in water and when out of water. In the latter method, Archimedes' principle states that the apparent loss of weight of the submerged object equals the weight of water displaced, from which easily follows the volume. The water submersion methods are precise enough, but when applied to persons, especially children and infants, they require, as charmingly understated by Greg Deskins, "a significant amount of subject cooperation."

The jug and the soft drink bottle, when driven to make sound in the way mentioned above, are properly called *Helmholtz resonators*.[4] Any volume of gas in an enclosure that has a single, restricted opening to the outside is such a resonator. The action is different from that of the closed or open-ended pipe, where the resonant frequency is found from the number of whole, half, or quarter wavelengths in the length of the pipe. In the Helmholtz resonator, the frequency depends on the volume of the enclosure and the area and length of the opening or neck. The frequency decreases if the neck decreases in area, or increases in length, or if the volume increases. In the typical case (e.g., a jug), the resonant frequency is so low that the greatest dimension of the jug is a small fraction of the wavelength of the sound. Consequently, the alternate compression and rarefaction in the air in the jug is nearly uniform and in phase, throughout the volume. The rapid movement of the air occurs in the neck. A through D in Fig. 1 show four stages of the cycle.

To digress for a moment, one can get striking effects by connecting a small loudspeaker to a variable frequency audio generator and using it to test a few common containers. The openings or necks of the containers are held near the loudspeaker, and resonance is recognized just by listening. I found that a beer bottle resonated at 200 Hz, and a 7-Up can at 330 Hz. The jug (Fig. 1E) gave especially strong resonant responses from ambient sounds. Its capacity is 1.4 ℓ and its minimum neck diameter is 6.5 cm. It resonated at 290 Hz. When the mouth of the jug was held near the ear in the presence of a piano solo (from tape), its response was heard only when a note near 290 Hz occurred. When in the presence of traffic noises or a passing airplane, it responded continuously, picking out its 290 Hz frequency from the full spectrum of noise. I did not succeed in getting a resonance in my mouth cavity—a disappointment because, as I have understood, it was that application that led Helmholtz to investigate the effect that now carries his name.

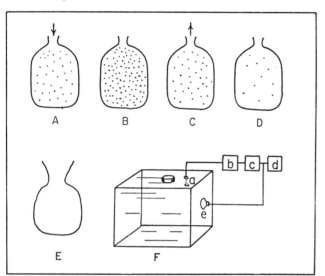

Fig. 1. A–D: A jug acting as a Helmholtz resonator. The motion and pressure of the air are indicated for four stages in a cycle. A. Velocity in the neck is maximum inward; pressure in the jug is neutral (just at the point of changing from rarefaction to compression). B. Compression is maximum, velocity in the neck is zero (just reversing). C. Velocity in neck is maximum outward; pressure is neutral (just between compression and rarefaction). D. Velocity in neck is zero (just reversing); rarefaction is at maximum. E. The glass jug with wide flaring mouth that gave exceptional response to ambient noises. Volume, 1.4 ℓ, smallest diameter of neck 6.5 cm, resonant frequency 290 Hz. F. Simplified diagram of the system that measures the resonant frequency in the plastic box that can contain a living subject. The opening corresponding to the neck of the jug can be seen at the top. a) microphone in the box. b) phase adjuster. c) amplifier with feedback for level control. d) frequency measuring unit. e) loudspeaker.

The resonator developed by Deskins and coworkers is a box of heavy transparent plastic, with a door in the side to admit the subject to be tested. One designed for infants has a volume of 100 ℓ, and one for adults and has a volume of 425 ℓ. Each has an opening to the outside (corresponding to the neck of the jug) of such area and length as to put the resonant frequency in the range of 20 to 25 Hz for the small box and about 15 Hz for the large one. Such low frequencies make the

wavelength of the sound long compared to any dimension of the enclosure (the order of 20 times), thus ensuring that the instantaneous sound pressure will be nearly uniform throughout the volume. The result, therefore, will not be influenced appreciably by the location or orientation of the subject being measured (as long as it is not close to the hole). To calibrate the system for resonant frequency vs. volume of the subject, containers filled with known volumes of water were used.

The precise measurement of the resonant frequency, which is the heart of the problem, is done in a way that is simple in concept and, therefore, pleasing to a physicist. A simplified diagram is shown in Fig. 1F. A microphone in the enclosure senses the sound; the sound is amplified and fed back into the enclosure by a small loudspeaker in one of the walls. There is always enough ambient noise to get the feedback started. Sound at the resonant frequency of the enclosure builds up, just as it does in an auditorium when the public address system is set with too much amplification. But there is a difference; a level-limiting part of the feedback loop holds the amplitude at such a value that distortion does not occur. A branch line from the feedback loop to a frequency-measuring device completes the design, as indicated in the figure.

The "bottom line" that is sought using the foregoing measurements is average density. Getting the volume is the tricky part. The average density then follows as the ratio of mass (this only requires a simple weighing with a known g value) and the volume. The result depends mainly on the relative amounts of fat, water, protein, and bone, listed in the order of ascending density. The range of average density from a very obese to a very thin adult is about 1.00 to 1.09 times the density of water.[5] Of course, four unknowns cannot be found from the one result of average density. Nevertheless, interpretations can be made that are valuable clinically.

Finally, let me mention two facts of physics that are important in the question of the precision with which the subject's volume can be found from the frequency. 1) The frequency tells the unoccupied volume in the box. If that (typically) is 5 times the volume of the subject, the error in the latter has an adverse factor of 5. 2) The fact that the frequency varies as the inverse square root of the volume brings another adverse factor of 2 in the error in translating from frequency to volume. Fortunately, of all the quantities in physics, frequency can be measured with relative accuracy and ease.

References
1. 11381 Meadowglen, Ste. I, Houston, TX 77082-2632.
2. W.G. Deskins, D.C. Winter, H.P. Sheng, and C. Garza, J. *Biomechanical Engr.*, Vol. 107, Nov. 1985.

3. W. Gregory Deskins, *Amer. Soc. Mechanical Engr.*, publication 86-WA/NCA-24, talk at Annual ASME meeting, 12/7–12/86.
4. H.L.F. Hemholtz, *Crelle* **62** 1–72 (1860). For a modern treatment, see *Fundamentals of Acoustics*, by Kinsler et al.
(John Wiley & Sons, 1978).
5. For more details see: J. Brozek, F. Grande, J.T. Anderson, and A. Keys, *Annals*, New York Acad. Sci. **110**, Part 1, 113–40 (1963).

A Tornado in a Soda Bottle
and Angular Momentum in the Washbasin

On sale now (and very popular) at the Ann Arbor Hands-On Museum,[1] and no doubt in gift shops in other places, are plastic pipe couplings, 5 cm long and 3 cm in diam, with inside threads in both ends.[2] Why do people buy them at $1.50? Because the threads are special: They will screw onto the 2-liter size plastic soft drink bottles. Two such bottles (preferably the colorless ones) can be connected together, like an hourglass (Fig. 1A). If enough water is included to fill one of them about three-quarters full, vortex effects can be made that intrigue not only the physicist but also the layman.

To make a vortex, the connected bottles are inverted so the water is in the upper one, then the top is moved rapidly in a small circle to produce a rotation in the water. If then nothing happens, mild shaking will cause a little water to go down through the necks in exchange for an air bubble, and that will start the continuing flow. A vortex will soon develop (Fig. 1B). If the initial rotation is sufficient, the vortex will be stable and continue until the upper bottle is empty, which will take about 30 s. If the initial rotation is not sufficient, the duct will pinch closed at some places and cause the air to come through as bubbles (Fig. 1C).

While the above observations satisfy the layman, the physicist wants to know more. If small bits of toilet tissue are put into the water, the movement can be studied (Fig. 1D). The tissue shows that for a good vortex to develop, the water must be rotating initially at 1 to 2 rev/s. In the fully developed vortex, the bits of paper near the center go too fast for the eye to follow. But from the law of conservation of angular momentum, an estimate of their angular velocity can be made. The estimate can only be an upper limit, because through friction some of the angular

momentum of the water is lost to the bottle and, in turn, to the outside world—the Earth. Nevertheless, the upper limit is quite surprising. We can estimate it from the fact that all of the water, before leaving the upper bottle, has to rotate within the neck of the bottle. We might think of a sample of water in the bottle (which has a 5.4-cm radius) in the form of a thin cylinder or 'pancake' rotating at 1 rev/s. When the sample descends into the neck its radius will decrease to 1.1 cm, and its thickness will increase accordingly (constant volume). Its moment of inertia will decrease by the square of 1.1/5.4 (neglecting the small duct in the center), and the angular velocity and revolutions per second will change by the inverse factor, to 24 rev/s. That's a "best case" (the rate of rotation certainly will not be that great), but it conveys the idea.

The coupling between the bottles has a constriction

Fig. 1(A–B). A. Two 2-liter soft drink bottles attached with the coupling; one filled with the right amount of water for the experiment. Explanation of the black caps: Typically the 2-liter bottle is made of thin plastic, round on the bottom. It has a heavier, flat-bottomed boot, cemented on, so it will stand upright. B. Vortex with a clear duct all the way, resulting from an initial rate of rotation between 1 and 2 rev/s.

in the middle that has only a 4.8-mm radius (probably arrived at experimentally, to maximize the lifetime of the vortex). If we do the same calculation for that radius, we get 127 rev/s! Of course it is not that great, friction having exacted its toll. Unfortunately the coupling is not transparent; if it were, we might see the duct of the vortex as a mere thread.

While thinking physics, we should wonder about the energy. When the radius of rotation of any object is decreased while conserving angular momentum, the kinetic energy increases. The extra energy has to be supplied. Swinging a stone in a circle on the end of a string is the familiar example; the energy is supplied by the force multiplied by the distance in pulling in the string. The analogy does not apply in the case of the bottles, where the only source of energy is gravity (the change in potential energy, mgh, from about the middle of the bottle down to the neck). You may wish to calculate the rate of rotation in the neck that could result from the energy derived from the fall from the middle of the bottle to the neck.

The observations with the bottles call to mind what almost amounts to an "old wives' tale," namely that water in a washbasin, if quiet enough initially, tends to whirl down the drain counterclockwise in the northern hemisphere and clockwise in the southern. The literature that accompanies the bottle-coupling seems to encourage that idea: "In seemingly perfectly still liquids even the Earth's rotation tends to determine the direction, counterclockwise north of the equator...." The fallacy of the washbasin story is in thinking that the water is going to whirl visibly one way or the other as it goes down the hole and that the small initial Earth-rotation just determines, or triggers, which way it gets started. It is true that as the water moves from the periphery to the drain hole the angular velocity does increase, but only by some factor (maybe 20) as a result of the conservation of angular momentum, as we discussed relative to the bottles. But if the water is quiet, except for the initial Earth rotation, the factor 20 will not make the rotation visible at the drain.

Is it possible to detect the rotation of the Earth in a washbasin-type experiment? No reason in principle it should not be–but not by eye. The initial rotation of the quiet water in the basin is, at 45° latitude, about 10°/hr.[3] If that is multiplied by 20 at the drain hole, it is about 3°/min. The water is gone in less than a minute, however, so the best that could be observed would be that a small marker floating on the water would approach the drain in a gentle spiral-like path, differing just a little from a straight line. It might be possible to photograph the floating marker with a time exposure. The rate of rotation at the hole might be increased by making the hole smaller. Great care would have to be taken to eliminate temperature gradients which could produce convective movement before the start of the experiment.

A question the bottle observation raises: Is the vertical component of motion in a real atmospheric tornado downward, as in the bottle, or upward? One has the *impression* that objects are picked up and carried aloft during a tornado.

References

1. P.O. Box 8163, Ann Arbor, MI 48107.

2. "Tornado Tube" by Burnham Associates, 26 Dearborn St., Salem, MA 01970.

3. The observable rate of rotation of the laboratory frame of reference is the rate of rotation of the Earth on its axis multiplied by the sine of the latitude. That appears clearly in the rate of turning of a Foucault pendulum.

Fig. (C-D). C. Vortex resulting from insufficient initial rotation, pinching closed at some points, causing the air to come through as bubbles. D. Pieces of toilet tissue in the water will show the motion (but not here, because the photographs were taken with a flash).

Uncommon Uses of the Stereoscope

Longer ago than I like to think, in a store in Central City, Colorado, I was thumbing through a box of old stereo cards—the kind viewed in Grandpa's stereoscope (Fig. 1). To my surprise, one showing the full moon turned up. A quick examination of the positions of the craters near the edges showed that the views were indeed stereo—that they had been taken from slightly different angles. (For 10 cents it was mine.) The stereo card showed that, contrary to the general impression, the moon does not at all times show its exact same face to us. Much later, quantitative answers to the matter appeared in an excellent article by Tom Greenslade.[1]

The moon appears to execute a slight rocking motion (more properly called *libration*), and that is what makes it possible to get stereo pairs of pictures. The libration has two principal causes. One is the inclination of the moon's axis of rotation with respect to the plane of its orbit around the Earth. Rotation? From our viewpoint the moon does not rotate. But in keeping nearly the same face toward us as it goes around the Earth, the moon has to rotate one revolution during each journey around. The axis of that rotation is fixed in the cosmic inertial frame. So as we view it, the face "nods" up and down (a north–south libration) about 6.5° each way. It repeats each month, but to see a full moon at the two extremes you have to wait half a year (Fig. 2A). You can see for yourself using an apple and a pencil.

The other principal cause of apparent rocking motion lies in the fact that the moon's orbit around the Earth is a little bit elliptical. The orbital angular velocity therefore is not quite constant. But the angular velocity of the moon on its own axis is constant. The combined effect, as we view it, is a rocking sideways, as if saying "no" (an east–west libration). That is about 7.75° each way.

A lesser cause of apparent rocking is our own travel on the rotating Earth between moonrise and moonset. That travel is some fraction (depending on our latitude) of the earth's diameter. It can give up to about 1° each way, as an east–west libration.

Getting a stereo pair of pictures that shows the moon in correct orientation is not just a matter of getting

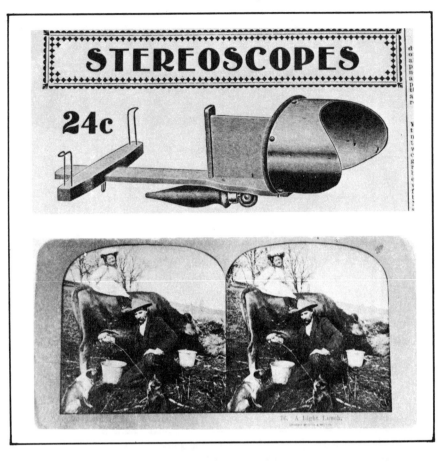

Fig. 1. (Upper) The Holmes stereoscope as it appeared in the Sears, Roebuck catalog, early 1900s. (Lower) One of the viewing cards from about the same time. The stream of milk from cow to kitten, in relation to the pail, shows the stereo effect (parallax). Millions of such cards were made, for entertainment and enlightenment.

pictures in two aspects that are different. For correct stereo effect when mounted on the card, the moon should appear turned a little from one picture to the other around an axis perpendicular to the line joining the eyes of the viewer, and therefore perpendicular to the long dimension of the card. Add to that the necessary condition that the moon be in the same phase, preferably full, in the two pictures, and predicting the times for getting those becomes a job for an astronomer! But there may be an easier (lazier) way. Just take a picture of each full moon through the year and pick the best pair. If you have a small telescope and can take pictures through it, you might try that.

The good luck at Central City got me started playing with the device with which the cards are viewed—the stereoscope—and putting it to some unconventional uses. Fig. 1 shows how it appeared in the Sears, Roebuck catalog in the very early 1900s. That model is attributed to Oliver Wendell Holmes. He didn't invent the stereoscope, but he put it into a form that became widely popular for home entertainment.[2] The

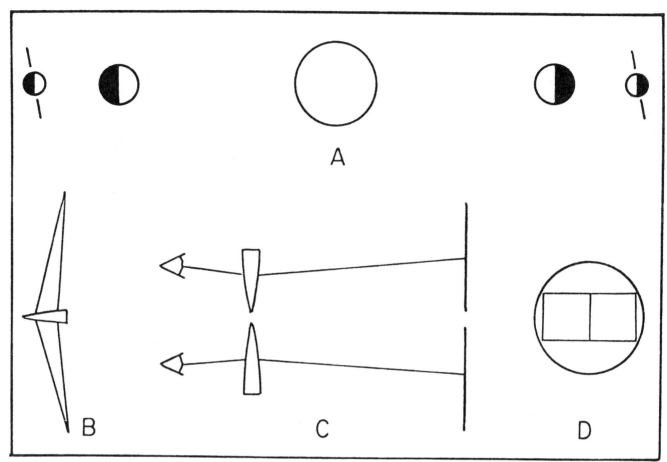

Fig. 2.(A) The sun (center circle) with the Earth and moon at times half a year apart, at full moon, and showing the maximum change in tilt of the moon's axis as seen from the Earth. (B) Object and image points for a piece of a spherical lens, showing the offset. (C) Arrangement of the stereoscope, showing lines of sight converging from the eyes to the lenses, then diverging to the corresponding points in the stereo pair. (D) How the lenses in the Holmes stereoscope apparently were cut from a 10-cm-diam, 20-cm focal length spherical lens.

stereoscope, as well as cards on every thinkable subject (Fig. 1), can still be found at antique sales. But not at the price shown in the figure!

The stereoscope makes use of an interesting property of a spherical lens. If the lens is cut or broken into pieces, every piece will project an image with the focal length that the whole lens had. And, because the pieces are wedge shaped (except for the center one), the image is offset (Fig. 2B). To reproduce in the simplest way the old-time stereoscope, all one has to do is break a lens in two and interchange the pieces, as in Fig. 2C. The pieces act as ''magnifiers'' to allow the eyes to focus on the card at a short distance, and they give a comfortable convergence to the lines of sight from the two eyes, even though the pictures on the card are further apart than the eyes. Central ray traces are shown from the eyes to points on the card, the latter points corresponding in the two pictures.

For anyone inclined to experiment, here are data found by measuring a Holmes stereoscope. The lens pieces have 20 cm focal length, evidently cut from a lens of about 10 cm diam and trimmed to 4 cm square (Fig. 2D). Their centers are at the intraocular distance, about 6 cm. The card is 9 × 18 cm, and corresponding points in the pictures are 8 cm apart. The holder for the card can be moved outward or inward on a track to suit the individual's focus and convergence. The holder and track are visible in Fig. 1.

Here are some things you can do with stereo pairs. You can make your own cards. Standard prints from color negatives are of ample size to allow trimming and straightening for mounting. Shoot one, step to the side and shoot the other (the subject holding the smile). You can do something the yes cannot do: In effect increase the intraocular distance—which here is the separation of the two camera positions—to maintain or enhance the stereo effect for distant objects. That begins to pay even for objects at a few meters. Fred Z. Hendel of our Physics Department showed me an extreme example of enhancing the stereo effect. While in a small airplane flying on a path parallel to, and to the west of the Andes Mountains, he took a pair of pictures at points several miles apart. When seen in the stereo viewer the impression of depth was striking. The Andes looked tabletop size.

It's interesting to reverse the stereo effect. I interchanged the pair of moon views. Presto! I was inside looking out. Even the craters were inside out. That trick works well only for a simple surface, having no part behind another part, and against a featureless background. Try it for a person's face.

What if you make a card with two pictures that are identical—not stereo? Viewed with both eyes, the scene looks flat. But if one eye is closed, something interesting happens (with me at least). There seems to be more depth, more like 3-D. It's understandable. In looking into the stereo viewer, two effects contribute most strongly to the illusion of depth. One is parallax: the difference in the pictures due to the different locations of the camera. The other is perspective: the diminishing apparent size of objects, the greater their distance was from the camera. In viewing the pair of identical pictures, the absence of parallax tells you that what you see is flat, while the perspective tells you it has depth. With one eye shut, you get rid of the conflicting signals, and the perspective alone does a surpisingly good job of giving the illusion of 3-D. You can try that with a picture in a magazine if it is one that has plenty of perspective. See if it works for you.

References

1. Thomas B. Greenslade, Jr., "The first stereoscope pictures of the moon," Am. J. Phys. **40**, 536 (1972).
2. The history of the stereoscope as well as many effects produced by it are treated in "Experiments with stereoscope images" by Thomas B. Greenslade, Jr. and Merritt W. Green III, Phys. Teach. **11**, 215 (1973).

A Magnetic Compass with No Moving Parts

Columbus used a magnetic compass on his voyage to the new world in 1492. Magellan embarked on his round-the-world voyage in 1519 under guidance of a compass.[1] The compass of that day was simply a needle on a pivot, the needle having been magnetized by rubbing it on a lodestone. Except for a few mechanical improvements, that instrument remained the same through the centuries until recent times. Despite its simplicity, it was an excellent indicator of magnetic direction for ships on the high seas as long as ships were constructed of wood. Problems arose as the wood was replaced by steel. The first solution was to locate the compass needle as far away from the bulk of the steel as possible, for instance on a high mast. The direction that the needle was pointing was sensed by a photoelectric eye or other means, and the information was transmitted by wire to a readout device in front of the pilot's eyes. This system works, but it has some problems stemming from the necessity of having the delicate compass needle unattended in a remote place.

Recently a way has been developed of sensing the direction of the Earth's magnetic field without the use of any moving parts. One benefit is that the sensing element is rugged, capable of being located in any remote place. Its output is transmit-

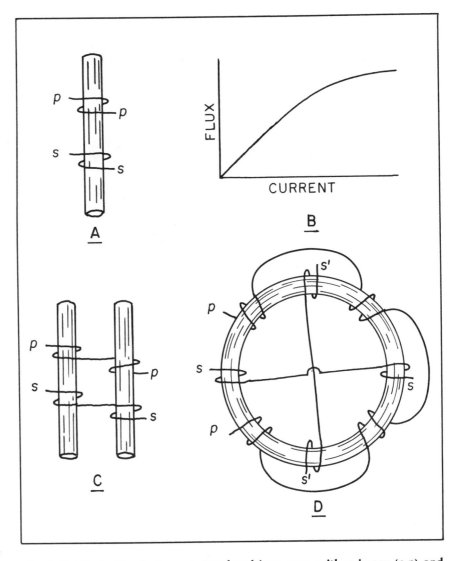

Fig. 1 (A–D). A. Simple iron or powdered iron core, with primary (p,p) and secondary (s,s) windings. **B.** Qualitative curve showing how the flux in the core changes with current in the primary as saturation of the core is approached. **C.** Two identical cores with primaries in series and connected so that the directions of flux are opposite. The secondaries also are in series, connected so that the emf's induced by changes in primary current subtract. **D.** Four of the transformers (as in A) joined into a ring forming a toroid. The primaries are connected so that the flux direction is everywhere either clockwise or counterclockwise. Two pairs of secondaries (s,s and s′,s′) produce difference-voltages that give the two 90° components of the external (Earth's) field. The directions of winding can be followed in the figure.

ted by wire to a display or readout conveniently located for the pilot. Such compasses are made by more than one company; my information was supplied by Robert W. Kits van Heyningen of KVH Industries, Inc.,[2] whose device is called Fluxgate.

The sensor works on an elementary physics principle: the generation of a voltage (more properly, the induction of an electromotive force, emf) in a coil by a time rate of change of magnetic flux that links it. But hold on: If the flux is that of the Earth's field, which is constant, and the coil is not allowed to move, how can a voltage be generated? There must be more to it. The solution is that the coil has an iron core, and clever use is made of the saturation properties of the iron. This can be explained in a series of steps (keeping in mind that they will be somewhat simplified, compared to the way it is done in the commercial devices).

Figure 1A shows an iron core with primary and secondary coils p,p and s,s wound on it—a simple transformer. The way the magnetic flux in the core changes with the current in the primary is shown in Fig. 1B. The

Fig. 2. A lodestone compass, in the form of a desk ornament. The piece of lodestone hangs from a Ping-Pong ball, which floats in water. It is kept centered by a fine steel wire that runs up through a hole in the jar lid to a N–S arrow. Friction is extremely low; the device will point if almost any piece of iron is substituted for the lodestone. A small magnet on a chain provides relief for the teacher between bluebooks.

leveling off of the curve at the high end shows what saturation means. When the primary current changes, emf is induced in the secondary in proportion to the *rate of change* of the flux that links it. However, as seen in Fig. 1B, a given rate of change of current produces a different rate of change of flux, and therefore, of emf in the secondary depending on where it is on the curve. It is a matter of the slope of the curve.

Now look at two of the transformers (Fig. 1C), identical in every respect. The primaries are in series, so that the flux, as well as the rate of change of flux, in the two cores will at all times be the same. They are connected so that when the flux is in the up direction in one it is in the down direction in the other. The secondaries are also in series, connected so that the voltage output of the two is their voltage difference. The directions of the windings and connections can be followed in the figure. Thus, the voltage output of the secondary combination will be zero at all times no matter how the primary current changes.

Now think what happens in the two cores if a small external magnetic field (the Earth's) is present, say in the direction from top to bottom of the page. That

external field introduces a little flux in addition to the flux that is present due to the primary current. Because the flux from the external field is in the same direction in both cores, it increases the total flux in one and decreases it in the other. So, at a given value of current, it puts one core a little closer to saturation and the other a little further from saturation—different places on the curve, where the slopes are different. Therefore, for any rate of current change, the secondary emfs are not the same; there is a net voltage output from the two in series. Thus, we have a signal to work with that indicates the presence of an outside field—the Earth's.

Of course, the Earth's field will be at some angle to the pair of cores, so the signal tells only the *component* parallel to the cores. To get the *direction* of the Earth's field, the component at right angles is needed, thus another pair of cores (not shown) is installed at 90° to the first pair. The four primaries connected are in series. The secondaries are paired, giving two separate components.

This description has explained how the basic information is obtained, without the use of moving parts. Now there are various options as to how the device is constructed physically, how the primaries of the transformers are driven, and how the signals are processed to give the readout to the ship's pilot. All of that comes under the heading of "state of the art." Some features will be of interest; what will be said will pertain mainly to the Fluxgate.

The four cores of Fig. 1C are combined into a ring, arranged so the saturation flux made by the primaries goes around in a circle. That arrangement makes a *toroid* out of it (Fig. 1D). The secondaries (s,s and s′s′) are paired as before, giving two components. The primary current is ac and of course high enough for near-saturation. The whole secondary (difference) output is not used. The second harmonic is selected out and used for the processing. The final result is displayed to the ship's pilot in various ways, one being an analog device that resembles a conventional compass. With the location of the sensor in the most favorable place, the problem of the effect of the ship's steel is not entirely escaped. Some coils or magnets still must be used to balance out residual fields. But there is an added degree of flexibility: The microcomputer that processes the signals can be programmed to eliminate certain errors of local origin.

For fun, look at Fig. 2, which shows the most primitive compass form: a piece of lodestone supported by water. It is almost certain that the earliest compasses were made by the Chinese in the 11th century A.D.[3], and that they consisted of lodestones floated in miniature boats in basins of water. The compass shown here is authentic in its use of a lodestone[4] and water, but it draws on modern technology for a Ping-Pong ball and a mayonnaise jar. The lodestone hangs from the Ping-Pong ball, and the wire running up through a hole in the jar lid keeps the ball centered. The only friction is in the contact of the wire with the edges of the hole, and that is remarkably small. So small, in fact, that if the lodestone is replaced by any piece of iron, such as a bolt, it will point, although taking longer to come to equilibrium. Of several pieces of iron picked out of the scrap box, all had enough permanent magnetism to work. The ancient Chinese need not have gone to the effort of finding lodestone!

If you want a conversation-piece for your desk, and you use mayonnaise and play Ping-Pong, try making a compass.

References

1. H.L. Hitchins and W.E. May, *From Lodestone to Gyro-compass* (Hutchinsons's Pub., London, 1952).
2. 850 Aquidneck Ave., Middletown, RI 02840.
3. Joseph Needham, *Science and Civilization in China*, Vol 4, pt 1, physics, (Cambridge University Press, 1962).
4. Lodestone is available from Ward's Natural Science Establishment, Inc. P.O. Box 1712, Rochester, NY 14603.

The Doppler Ball: A Novel Use of the Piezo-Electric Effect

I thought the Doppler effect might make a good subject for "How Things Work" when Warren Smith showed me an unusual demonstration of it. Warren is in charge of apparatus and demonstrations for the introductory lectures in our department. He built a solid-state sound generator into a plastic Whiffle ball. Playing catch gives a substantial frequency shift up and down, as is heard by the thrower and catcher. He said the idea was not his, so a little tracing was in order. I found that Frank Crawford of the University of California at Berkeley had long ago described the forerunner of the demonstration—a bare battery and a soundmaker taped together, thrown back and forth by players wearing baseball mitts.[1] He called it a "Doppler ball" although it was not really a ball.

I was intrigued enough to put together the Warren Smith version. The soundmaker is a "piezo allerting buzzer" from Radio Shack.[2] It produces a pure 2800 Hz tone, and when powered from a 9-V battery it can be heard at about 30 m; with two 9-V batteries, 43 m (my test). It can be made even louder, standing up to 28 V. The ball is made of unbreakable plastic from Toys-R-Us, 10 cm in diameter. It must be sawed in half and put back together with plastic tape. The ball has open slots, so there is no problem with the sound getting out or making an on–off switch accessible. Fig. 1 shows the equipment with padding and the assembled ball.

The frequency shift to be expected is ample for hearing the effect, as can be easily calculated.[3] We know the frequency (2800 Hz) and the speed of sound (344 m/s), but we need the speed of the ball. Say it is thrown to a catcher 30 m away at an initial angle of 30° from the horizontal. Neglecting air drag, the velocity at the beginning and the end of the trajectory is 18.4 m/s. The thrower hears 2657 Hz as the ball leaves, and the catcher hears 2958 Hz as it arrives, a difference of about a full note on the musical scale. What the catcher hears as the ball leaves the thrower and vice versa would be

Fig. 1. (left) The Doppler ball, open, showing the piezo buzzer, battery, switch, and padding; and (right) closed with tape ready for a game of catch.

interesting to calculate (not forgetting the 30° angle).

In the buzzer the piezo-electric effect is used in an unusual way. Anyone familiar with the typical uses of piezo crystals[4] will think right away of one vibrating in the frequency range of megahertz and with very small amplitude. The puzzling questions are: (a) How is the frequency of vibration brought down to the audio range, and (b) how is the amplitude made so large as to produce the loud sound? I found some answers in a British newsletter.[5] (It saved me from having to dissect my buzzer!) There were two surprises: first, the crystal is in the form of a diaphragm, and second, the chamber above it makes use of the Helmholtz resonance.[6] The cross section, as described in the article, is shown in Fig. 2A. Incidentally, it is described there as the alerting element of a smoke alarm, called a "horn"—a more appropriate term than "buzzer". But it is the same device.

The piezo crystal is a thin disk, metallized on both faces for electrical contact. It is attached (probably cemented) to a thin metal disk, which is clamped against a circular knife edge. The combination can vibrate as a diaphragm or drumhead, as shown much exaggerated in Fig. 2B. In the diaphragm mode of vibration its natural frequency is about 2800 Hz. The chamber above it is a Helmholtz resonator, the outside area of the opening adjusted so that the air and the diaphragm act as a system resonating at 2800 Hz. The system maintains a high amplitude, emitting sound through the opening on a suprisingly small electrical power input that measures 36 mW for operation on 9 V. The driving circuit is contained in an IC chip which is potted in wax and cannot be viewed (nondestructively).

One cannot help comparing the above system with a singer: The lips and mouth cavity are formed to act as a Helmholtz resonator driven by a vibrating element. The only difference is that the latter is not a piezo crystal!

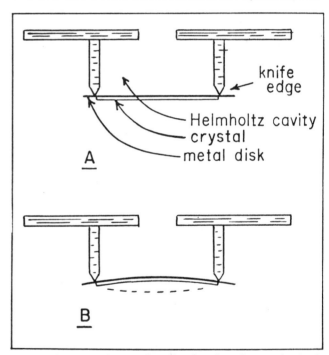

Fig. 2. The piezo "horn" after the sketch in the article cited. (A) The diaphragm combination at one end of the Helmholtz cavity and the (carefully adjusted) restricted opening at the other end. (B) The (greatly exaggerated) drumhead mode of vibration of the crystal-metal disk combination.

It is not easy to see how the piezo effect drives the diaphragm or drumhead mode of vibration. It must happen in the following way. The electrical potential difference is applied on the two faces of the crystal disk; therefore, its thickness dimension must be the one that increases at 2800 Hz. That makes the plane containing the center of mass of the crystal move toward and away from the diaphragm at the frequency, and that drives the drumhead resonant vibration of the combination. (This is my interpretation; it was not given in the article cited.)

The piezo buzzer should be among the tools of trade of every lecturer. Due to the combination of the resonant diaphragm and the "tuned" Helmholtz cavity the sound is remarkably free of overtones—the output is nearly pure sine wave. For that reason, the device is excellent for demonstrating standing waves in a room. And for the more adventurous, how about mounting the device in the underside of a Frisbee?

References

1. Am. J. Phys., **41**, 727 (1973).
2. Several versions are listed by Radio Shack. The one used here is #723-068, $7.95. Diaphragms alone are offered by Edmund Scientific. 1-in.-diameter, thin metal disk with piezo crystal attached. 6 for $5.95, #A35,200 in the 1987 Annual Reference catalog.
3. The Doppler formula is found in most texts, but is given here for convenience. Source moving toward stationary observer: $f' = fV/(V - v)$. Source moving away: $f' = fV/(V + v)$. f' is the frequency heard. f is the frequency of source. V is the velocity of sound. v is the velocity of source.

   The calculation of the initial velocity of the ball in flight over a trajectory is standard, involving no more than resolution into components and acceleration due to gravity.
4. If a reminder of the piezo-electric effect is needed: Some crystals, notably rochelle salt, tourmaline and quartz, exhibit the effect. If a potential difference is applied to two opposite faces (properly chosen with respect to the crystal axes), a dimensional change occurs. Conversely, if a force is applied to produce the dimensional change, a potential difference between the faces appears. Furthermore, the crystal has a natural frequency of mechanical vibration, such that the two faces mentioned move toward and away from each other with a nodal-plane halfway between.

   If the applied potential difference is alternating at the crystal's natural frequency, vibration will be driven. Since potential difference produces mechanical change and vice versa, a simple feedback loop through a transistor makes the vibration self-sustaining. The resulting frequency, determined only by the physical properties of the crystal, is highly stable and permanent.

   Such crystal oscillators are widely used, for example, to fix the frequency in a radio transmitter or a watch. Quartz is the most used because of its durability and small change in frequency with temperature. The pieces used are wafers from less-than-a-mm to a few-mm-thick, cut from larger crystals. Their frequencies, in the fundamental mode, are in the range 1 to 10 MHz. Where a higher frequency is needed, the wafer is made to vibrate in an overtone mode. Where a lower frequency is required, (e.g., 1 Hz to move the second hand of a watch) the crystal frequency is "scaled down" in an IC chip.
5. Snippets, #4, Autumn 1983. An educational newsletter by the Institute of Physics, 47 Belgrave Square, London SW1X 8QX.
6. Resonant vibration of the air in an enclosure having a restricted opening to the outside (e.g., a jug). Examples were described in "How Things Work" in Phys. Teach. **25**, 454 (1987).

How to Rotate a Ball from the Inside

Our editor, Don Kirwan, called to my attention a ball that, while rolling on the ground, can be instructed by radio to go forward or backward and to turn right or left. He sent me an "operating manual" but no ball to go with it! The gadget is called a "Go Ball."[1] It is a plastic sphere, 18 cm in diameter, that is smooth on the outside—no hidden wheels, hooks, or air jets—only an on/off switch behind a small hole. The thought of a rolling ball changing direction all by itself makes one say "Hey, what about the conservation laws—momentum and energy?" Wouldn't a billiard player or bowler like to know that secret!

With a little tracking I located the inventors of the "Go Ball," Peter Green and Len Clark of Polygon Design.[2] They were generous in providing the needed technical information.

A good way to approach explaining the "Go Ball" might be to recall a simple example: a child's toy made of a tin can that reverses and comes back when rolled away on the floor. It is simple to make (Fig. 1A). A rubber band is connected between the ends of the can, on axis. A weight, such as a piece of lead, supported by a stiff wire hangs below the rubber band. Rolling the can away twists the rubber band so that the weight assumes the position in Fig. 1B. The center of mass of the whole toy is then to the right of the contact with the floor, so the can rolls back to you. Today one would use a mini, permanent-magnet motor (perhaps a geared-down one) instead of the rubber band, and let the battery be the weight supported below. But then it would not come back!

The "Go Ball" uses the same method as the self-propelled tin can for its forward and backward motion, but an additional problem has to be solved, that of steering (it is cleverly done). The cut-away in Don's operating manual shows the ball having an axle on a diameter of the sphere. By analogy with the Earth, the polar axis, the axis on which rotation will take place. The sketch in Fig. 1C shows the axle (a) as horizontal. It also shows an outline (b) indicating (without detail) the equipment (radio control circuitry, motor, and battery pack) supported on the axle so that most of the mass is below the axle. The radio antenna (c) sticks up into the otherwise empty hemisphere above the axle. While operating, gravity keeps the cluster of equipment below the axle, while power from its motor rotates the axle. The ends of the axle are fixed to the ball, so when the motor runs the ball rotates, but the equipment does not. That makes it roll. As shown, the center of mass of the equipment is below the center of the axle, so the axle is horizontal, thus the ball will go straight, rolling on its "equator" (d).

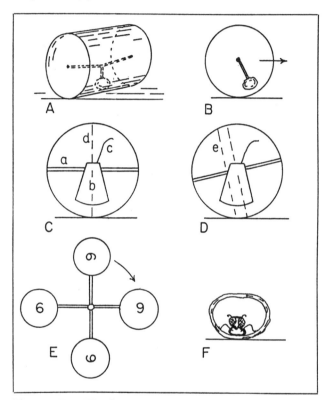

Fig. 1. A–F: A. The tin can that reverses and comes back. B. The tin can, showing the position of the lead weight when the rubber is wound up. C. The "Go Ball." a, the axle; b, the equipment package; c, the antenna; d, the "equator" on which it will roll. D. The axis tilted so that the ball rolls on a circle of latitude, e, and therefore in a curved path. E. Perpetual motion machine, adapted from the tin can/rubber band. F. The Mexican jumping bean—inhabited.

Now the tricky part—steering. To make the ball turn, a second motor acts to produce a tilt of the axle from the horizontal. It accomplishes this by moving some of the mass within the equipment package toward one end of the axle or the other, so that when the center of gravity is at minimum height, as always, the ball and axle repose as shown in Fig. 1D. Now the contact of the ball with the ground is not on its equator but on a circle of latitude (e). If we visualize the ball being rotated about

its polar axis by the first motor, while maintaining contact with the floor on the circle of latitude, we see it traveling in a curved path. Skeptical? I would be too, were it not for the fact that the ball exists and does go in a curve.

The motors in the "Go Ball" are controlled by what appears to be the standard radio control system used for model cars and boats. The control is proportional and reversible. The radio frequency is in the 27 MHz range. Balls using slightly different frequencies are offered, so that two of them can have a race or a battle. I tried to find out the maximum speed of the ball. Peter Green of Polygon Design came nearest to giving a number. He said "substantially faster than a walk." He went on to say that his company has built a 2-m-diam version that carries a passenger at 30 mph (13 m/s).

The idea of controlling something from the inside has interesting ramifications, worth a further look. For instance the rubber band device lends itself to a "perpetual motion" demonstration (Fig. 1E). Identical cans are mounted on the ends of four spokes. If the bearing at the center has sufficiently low friction and if the balance is precise, only one of the cans needs to have the rubber band and weight inside. Numbers on the cans (supposedly) represent the weights. The rubber band is wound up ahead of time by rotating the wheel backwards. When let go, it rotates because the weights are 9 (kg or lb) on the way down, and 6 on the way up. If you need something to break the ice for a lecture, you might try this demonstration.

An example of control from within that I have always liked and have written about is the lowly Mexican jumping bean.[3] A small "bug" invades a specific kind of Mexican bean and eats the inside so that as he grows he inhabits a hollow capsule (Fig. 1F). (We assume he knows how to drill out of it!) The bug jumps by applying a downward force to the capsule, which is of course transmitted to the table on which it rests. If he jumps hard enough to make impact with the "ceiling" of his hollow space, the bean is lifted a small distance off the table. Usually the bug does not know which way is up, so moves sideways, merely tipping or rocking the bean. Beans with live bugs are usually on sale in novelty stores on either side of the Mexican border.

In the above examples, it is easy enough to see how the conservation laws are satisfied: momentum or angular momentum is exchanged with the table or the ground. But go a step further: What if you are in a capsule in free space? Is any control from within then possible? Surprisingly, a certain kind of control is possible. Imagine that your capsule has only one window, and you want to change the orientation so you can look out in a different direction. You can do that by simply rotating yourself (by any means possible in your "weightless" condition) until the window comes in front of your eyes again. If the new view doesn't suit,

How Things Work

repeat. However you manage to turn, you impart angular momentum to the capsule equal and opposite to yours (the *total* being unchanged). When you stop turning, the capsule stops turning, but at a new orientation. The interesting part is that without the aid of any force from outside, you have permanently changed the capsule's and your orientations. You cannot choose any new direction of view you wish, because you stop turning only when your eyes are again in front of the window, the amount of turning is quantized! The principle would hold true in a real space ship, but you would have to rotate many revolutions to get a noticeable effect, because of the large ratio of the ship's moment of inertia to your own.

A similar example involving the whole Earth can be imagined. If a train goes from Detroit to Chicago, the time of sunrise for the whole earth is made earlier (certainly unmeasurably), and it stays earlier only to be undone by the train returning to Detroit!

References

1. The "Go Ball" is made in Japan. The U.S. distributor is Shinsei Corp., 12951 E. 166th St., Cerritos, CA 90701. It has been offered in *The Sharper Image* catalog, 650 Davis St., San Francisco, CA 94111, but currently is not available.
2. 616 Main St., Schwenksville, PA 19743.
3. *Phys. Teach.* **7**, 371 (1969).

Help for Forgetful Car Owners and Lazy Yo-Yo Throwers

The "CarFinder" and the Garage-door Opener

When searching for your car in a parking lot, having forgotten where you left it (as I do more than occasionally) you can press the button on a pocket-size radio transmitter, and your car will signal back to you. It will blow its horn, flash its lights, turn on its radio, or do anything else it has been programmed to do. For the details of how this is done, we thank Mark Gottlieb, who is president of the company that makes the device, called "CarFinder."[1]

The "CarFinder" is an offspring of the garage-door opener—one of those "why-didn't-I-think-of-it" ideas. The two devices employ the same principles of operation, so what will be said here about one system will apply to the other. The interesting physics concerns how the receiver is made to respond only to the radio signal from the owner's transmitter, never to surges from lightning, radio talk from passing taxis and police cars, or radio signals from would-be burglars. It's done by requiring that two frequencies, plus the time duration of the signal, be right—a combination hard to hit upon accidentally.

The transmitter of the "CarFinder" (the part held in the hand) sends a radio carrier wave, at a frequency of the *order of* 100 MHz, *amplitude modulated* at a frequency in the audio range (a few hundred to a few thousand Hz). The receiver, besides being tuned to receive the carrier, contains a filter that will pass only the modulation frequency. So, the receiver will not respond unless both frequencies are correct. Transmitters and receivers are sold in pairs, the frequencies having been set at the factory. The number of different combinations of carrier and modulation frequencies is great enough

to make interference between two devices highly unlikely.

The third safeguard is a time delay. The part of the circuit that closes the relay (that starts the door-opening motor or turns on the lights or horn) will not operate unless the received signal lasts for a second or more. You may have noticed that to open your garage door, you have to hold the button down, not just tap it. This prevents the door opening because of line pulses such as those from lightning.

Things were not so sophisticated in the earlier days of garage-door openers: There was no modulation, only the carrier. I owned that kind and can remember occasionally finding the garage door open. Now, a misfire can be due only to a passing taxi, with radio on at about the right carrier frequency, carrying a lady passenger practicing arias from the opera, who can hold the right high note for more than a second.

To be more specific about frequencies, the "CarFinder" uses a carrier in a range centered around 49 MHz, and any one of five modulation frequencies between 200 and 1500 Hz. Garage-door openers generally use higher frequencies: 200–300 MHz for the carrier and 10 kHz or more for the modulation. The power radiated by the transmitter of both devices is just under that requiring licensing. (Mark Gottlieb tells us that an electric field strength of 0.01 V/m at 3 m would require licensing.)

Finally, a suggestion to the maker of the "CarFinder": Why not have it unlock the doors, to save fumbling with keys?

The Automated Yo-yo

To shift to another invention that saves thinking, we turn to the yo-yo that automatically wakes up after "sleeping." In the common yo-yo, the string has a loop that is loose around the rod that connects the two halves of the yo-yo. If the yo-yo is expertly thrown, it will remain at the end of the string, spinning, the rod turning in the loop. A quick dip of the hand will allow a few turns of string to wind up, ridding the loop of slippage. The yo-yo climbs.

Fig. 1. "Yomega, the yo-yo with a brain."

Fig. 2. The separated yo-yo halves, showing the sleeve where the string is attached. The clutch that grips the sleeve is also visible.

In the automated version, named "Yomega,"[2] there is no loose loop; the string is fixed to a short plastic sleeve on the rod. The sleeve can remain stationary while the rod rotates. That allows the yo-yo to "sleep," while spinning at the end of the string. But it accomplishes no more than the loose loop did for the ordinary yo-yo. So what's different? This yo-yo contains a centrifugally activated clutch which, when the speed of rotation falls below a certain value, grips the sleeve so it will rotate with the yo-yo. It winds the string, making the yo-yo climb. A photograph of the yo-yo with the string wound in the space between the halves is shown in Fig. 1. The parts of the clutch are visible. Two plastic bars (they appear dark in the photograph) clamp the sleeve. Each bar is pivoted at one end and holds a steel ball at its other end, for added mass. The springs holding them in the clamped position are visible. At a certain speed of rotation the bars pivot outward and release the sleeve.

Fig. 2 shows the halves of the yo-yo separated. The sleeve, with string attached, appears white in the photo. It is in the grip of the clutch. Note that the clutch only occupies one of the yo-yo halves.

I am writing this because a subtle bit of physics puzzled me. When thrown hard, the yo-yo acquires a much greater speed of rotation than that which prompts the clutch to release the sleeve. Therefore, the yo-yo sleeps for a long time waiting for the speed of rotation to fall to the value at which the clutch regrips the sleeve and starts the string winding up. Why, during the "throw," doesn't the yo-yo start sleeping as soon as the rotation speed gets to the value that releases the clutch, instead of continuing to gain rotation speed? The answer was found by a simple experiment.

I first had to fix the clutch to be in the released condition when the yo-yo was not rotating. This was done by drilling two small holes at strategic places, through which pins could be inserted. After winding the string in the groove between the two yo-yo halves (as if ready for a throw), I inserted the pins to free the sleeve. Then I held the yo-yo and pulled hard on the string. The sleeve did not rotate and let the string pull out; instead, torque was exerted on the yo-yo. Then I began removing turns of string. The sleeve did not rotate, allowing the string to pull out until the number of turns had been reduced to four. Conclusion: The acceleration of the throw does not end until there are only about four turns left, even if the clutch releases earlier.

There is still the question of how the string grips the yo-yo while the sleeve is free. The width of the slot in which the string is wound is only about twice the width of the string. As long as there are more than a few turns, the string wedges in the slot, coupling firmly to the yo-yo without help from the sleeve.

References
1. Design Tech International Inc., 941-B 25th St. N.W., Washington, DC 20037.

2. "Yomega, the yo-yo with a brain" (about $10) Caffrey, Inc., 1641 N. Main St., Fall River, MA 02720.

What Can a Dimple Do for a Skipping Stone?

Spring is here, and the weather is right for skipping flat stones on water and making some observations of the physics involved. My interest in stone skipping, which led to some experiments, was rekindled by a package of manmade "stones" called "World Class" that I happened to see advertised and could not resist.[1] The package stated, "Our engineers discovered that a dimple in the center of the underside of the stone creates a wake on which it rides.... The result is the world's only hydrodynamically perfect skipping stone.... Incredible distances.... Use caution when throwing where swimmers, boats, and waterfowl are present." If we believe that the dimple does some good, it must be by improving the stone's "lift-to-drag" ratio during the time it is in contact with the water. If so, the loss of forward velocity at each skip is lessened, so the stone can make more skips before stopping.

Fig. 1 shows the stones in several aspects. They are 4.6 cm in diameter and 7 mm thick. The dimple is indicated by the circle on one of the stones. The surface from the circle, but only by about half a millimeter at the center, and the same at the outer edge. Not much of a dimple!

Stone skipping is serious business for some. Every summer a competition is held on Makinac Island in the Great Lakes, and the winning number of skips goes into the *Guinness Book of World Records*. Only the skips of natural stones are recognized; too bad for the makers of the dimpled stones, who claim 36 skips compared to the officially recorded number, currently 24. There is available literature on stone skipping,[2] and an ever-interested audience. A variant on the water skipping that has fascinated some experimenters is skipping a stone on the sand of a beach, right after it has been smoothed by a wave. At least it has the advantage that the stone can be retrieved! Skipping on both sand and water has been studied using flash photography.[3] Not all skipping has been for fun: Long ago in naval warfare, efforts were made to make projectiles skip along the ocean surface, the better to hit targets at the waterline.[4] And then there is the fabled throw of a silver dollar across the Potomac by George Washington. Was that done by skipping? I am indebted to Lewis Slack of the American Institute of Physics[5] for guiding me to several interesting sources.

What impresses anyone with an interest in physics is the remarkable stability shown by a skipping stone during flight. In spite of disturbances by rough water, differences in the contour of the undersurface (of natural stones), and variations in launching (sometimes poor), the stone quickly assumes an attitude with respect to the water surface that it maintains throughout the flight. It is almost parallel to the surface, front end a little raised. And it does not lose that attitude during its leaps through the air. The "planing" of a flat object on water is not unfamiliar, for example, a surfboard or a fast motorboat. They are stable: A tilt to one side increases the force of the water on that side and the tilt is corrected. The same correction occurs when the leading or trailing end gets too low or too high. But one would doubt that if either of these objects were to leap through the air a distance equal to 10 to 100 times its own length (which would be in scale with the skipping stone) it would come down in its same attitude! Lesser leaps than that have occurred in motorboat races, with disastrous results.

Fig. 1. Several of the manmade skipping stones. The circle on one shows the boundary of the dimple. The half-black stone equipped with a mast is ready for an observation of its spin and attitude with respect to the water surface.

So, we look to the spin of the stone to furnish the stabilization during its leaps. It becomes a miniature Frisbee! And that suggests two questions: (1) Is it really spinning at a high rate? Does the spin persist throughout the flight, or is it lost to the water during the brief contacts? (2) What about gyroscopic precession due to torques that might act on the stone when it is in contact with the water, so as to turn it, say edgewise?

I tried to satisfy my curiosity about the above two questions by doing a few simple experiments—not precise physics, but entertaining. For convenience, I used the manmade stones, but the results also should apply to natural ones.

First, I investigated the spin. Note in Fig. I that half of the top of a stone has been painted black. (Ignore the white stick for a moment.) When I skipped such stones, the top was a blur to the eye throughout the flight, showing that the spin was not quickly lost. That makes sense, because the stone is in contact for a very small fraction of time.

Next, what is the spin rate? That must vary from one thrower to another. To test my throwing, I tied a fine thread about one meter long to the mast (the white stick shown in the figure) so that it would trail behind and wind up on the mast as the stone rotated. I threw the stone into a snowbank about three meters away, so it could be retrieved. (My one wintertime experiment!) For several tries the result was two to three revolutions per meter. At 15 m/s, a reasonable throw, that would be 30–45 rev/s—as fast as most electric fans go.

What about gyroscopic precession, when the water is exerting a force on the stone? Certainly it happens: The question is whether it is in the sense that *corrects* deviations of the stone from its equilibrium attitude or in the sense that makes matters worse. In other words is it a *stabilizing* effect? I arrive at the answer "yes," but you should check me. (To agree on signs, say the stone is thrown by a right-handed person, therefore rotating clockwise when seen from above.) Suppose the leading edge dips too low, making the force of the water act effectively ahead of the center of mass. That makes a torque around an axis transverse to the direction of flight. Gyro action makes it tilt (I say counterclockwise) as seen by the thrower. That moves the force of the water to the left of the center of mass, causing a precession that raises the leading edge. So the deviation we started with is corrected. The logic works for an opposite deviation: You just change all the signs, and find that the deviation is corrected. How much of a part the gyro effect plays while the stone is in contact with the water has to be a guess.

One more use of the masts on the stones: They gave an idea of the attitude when skipping. They did not seem to wave around; they seemed to just lean a little backward and to the left. Maybe the stone finds a steady condition between the two gyro effects mentioned above.

With summer nearly here, try some stone skipping. You can't lose; remember that 24-skip goal.

References
1. Sun Products Group, 3402 A St. S.E., Auburn, WA 98002.
2. For example, a chapter in: James Trefil, A *Scientist at the Seashore* (Chas. Scribner, New York, 1985).
3. "The Amateur Scientist," *Sci. Am.* (Aug. 1968), pp. 112–118; (Apr. 1957), p. 185.
4. *Science 85* (Oct. 1985), pp. 86–87.
5. American Institute of Physics, 335 E. 45th St., New York, NY 10017 (recently retired).

Looking Backward and Forward— Start of a New Year

Having written (hard to believe) 46 "How Things Work" columns and looking a new TPT-year in the eye, the time may be right for comments as to where we have been in five years and where we may go from here—and most importantly, to renew your invitation to send suggestions and information. We like topics that contain worthwhile lessons in physics, and points not normally encountered in coursework, at least not in the same application. I will do most of the work and writing, but I need something to start from: The more information you send as a starter, the better. As the number of past columns grows, so does the chance that the subject you think of will be one already covered. To save you from that, the past topics are listed.[1]

Let me mention some of my exploits in retrieving the often oddball information for the columns, and let me persuade you that ferreting out new ideas to send to me might be fun and educational. For example, I did not know how gasoline pumps know when my tank is full, how cable cars get through the intersection of two tracks, why the bar codes on grocery items do not stop the grocer from changing the prices, or why new car batteries do not need distilled water.

Ploys of all kinds are necessary. The one I enjoyed most is what an engineer friend once dubbed "reverse engineering": dissecting the gadget (if possible without destroying it) and/or subjecting it to bench tests. For example, the touch buttons that call elevators—the kind that are activated by contact with the finger, rather than by being pushed. The first exercise was to remove the cover plate and probe around with a voltmeter; the second, to insulate my feet from the floor and see if the touch button would still be activated. It all had to be done without disabling the elevator. In the end the elevator repairman saved me from defeat by producing a sample of the critical part, the gas tube, which could be put to a bench test. The ever-spinning top yielded easily to tests with a voltmeter and ammeter. The plywood goose, first seen in a store in Canada (a Canadian goose, of course) seemed to seek an equilibrium configuration for no obvious reason. That was solved by experiments with a force diagram made of sticks and string.

Sometimes things fall into my lap, as happened while waiting for an airplane. A toy operating in the gift shop had penguins moving up an escalator while the escalator was not moving up. That secret cost $20; there was no other way but to take it home and open it up. The tornado in the soda bottle was easy to analyze, once bits of toilet paper were added to show the motion of the water. The "yo-yo with a brain" could not be opened short of complete destruction, but holes drilled into it solved the problem. And for fun in the open air, experiments with the manmade skipping stones were hard to beat.

The local environment can be a fertile source. I hit pay dirt at a sales and service establishment for bar code, checkout-counter equipment, the city parking and traffic department for the computerized traffic light system, and the police department for alcohol–blood level measurers. From local industries I found out about blood pumps for heart surgery and how ball bearings are made. Toy stores, gift shops, and hobby shops are always likely places—but you may have to buy the article before you can take it apart!

A source of an unexpected kind especially pleased me: a dusty volume in the basement of the library. It explained how houseflies maintain their equilibrium and maneuver away from fly swatters. In that case the reverse engineering had to come after the fact—that was to put a straight piece of steel wire in the chuck of a lathe, twang it, and slowly rotate the chuck. It worked just like a fly's haltere!

Suggestions I receive often concern security systems of one kind or another. Naturally intriguing subjects include: automatic bank tellers, money changing machines, gates to detect articles being smuggled out of libraries and stores, and coin testers in vending machines, telephones and parking meters. Knowing how these things work can be a step toward being able to defeat them for profit or for the challenge. The makers know that, and they are tight with information. I agree with the makers in some cases. For example, a reader sent me the security insert out of the cover of a library book. A little reverse engineering revealed how to take it through the checkout gate without setting off the alarm. Do not expect those instructions in this column.

Here is a trick you can no longer try because the technology has changed, but it illustrates the above point. It is how some of my fellow college students were able to make free telephone toll calls. They learned that the way the operator knew (in those days) what coins were put into the slot was just by hearing the ring of a bell via the transmitter. So they found a way to make their own 5-cent ding, 10-cent ding-ding, and 25-cent dong. Sounding it in front of the mouthpiece brought the operator's "thank you."

Fig. 1. **Views of the inside of the parking meter that has remained essentially the same for half a century. There are even more ratchets, levers, and cams than are visible in these two views. The meter does not even sort the coins: The customer does that by putting them in the right slots. A slot will not admit a coin if its diameter or thickness is too great. The mechanism inside will not give time if its diameter is too small. Those are the only tests of a coin (or slug) that are made. One spring, wound by the exterior "crank" supplies the energy to run the balance-wheel clock and lower and raise the "flags." As can be seen, the balance wheel is almost identical to that of a mechanical alarm clock.**

A subject whose time has come for this column is the electronic parking meter that has been turning up on the streets. It looks nearly the same from the outside as the ones that have been standard for half a century. But inside, the difference is that between a hand-held calculator and the old mechanical adding machine. Can you imagine the insides of a conventional parking meter being replaced by an IC chip? Probably not, because like everybody, you have seen the outsides of a hundred thousand meters but never the inside of one. To remedy that, Fig. I looks inside a vintage Chicago parking meter, when a nickel bought an hour of parking time. The question about the new meter that would interest us is how it uses electronic sensing to check the coins. (The interest is purely for physics, no intention of cheating the meters!) In the new meter there is only one slot, so the circuitry has to determine the denomination as well as test whether the coin is real or bogus. We can list the physical properties of a coin that might be tested: diameter, thickness, mass, specific resistivity, reflectivity, moment of inertia, and ferromagnetic response. Certainly it would not be necessary to check all of these properties. But which, and how? So far my efforts to find out have been like punching a pillow!

Some reader-suggested subjects await information or ideas. Maybe you can help with information on: wrist-watch size pulse-rate readers, touchtone and/or cordless telephones, room air "purifiers," laser printers, and compact disk players.

Our influence spreads. A friend purchased from a curio store in Carmel, California, one of the ever-spinning tops. Inside the package was an excerpt from this column, telling how it works (giving proper credit). Did the makers not understand it until TPT came along? The makers of the Tornado Tube are now quoting this column as to how that works. (Permission was given in both cases.)

I look forward to your input.

References
1. **1983:** Introduction, frisbees, elevator buttons, liquid crystals, gyrostabilized houseflies, everlasting light bulbs. **1984:** Metal locators, ever-spinning top, picking locks, blood pump, toasters, traffic signals, copy machines, smoke alarms, auto-focusing cameras. **1985:** Halogen lamps, wall-stud locator, ship locks, thermobile, spider webs, alcohol in breath, maintenance-free batteries, fluorescent lights, car air bags. **1986:** Ring interferometer, cable cars, plywood goose, cardiac pacemaker, spinning can and other, bar codes, car ignition, Aerobee and repelling deer, making steel balls. **1987:** Transponder, 3-D Moire', gasoline pumps, hydraulic ram, penguin escalator, body volume by sound, tornado in a bottle. **1988:** Stereoscope, nonmoving compass, Doppler ball, Go Ball, yo-yo with a brain, skipping stones.

Brrrr!
The Origin of the Wind Chill Factor

Wind chill figures will be heard in most radio newscasts this month—at least in the parts of the country that have a winter. If you have a computer and punch in the little program given here, you will be able to check the experts.

The figure given on the radio and in the paper is the *wind chill equivalent temperature*—often just called the wind chill. It is supposed to tell you the temperature it will feel like in the wind, at the currently measured speed and temperature. Everyone knows that what it feels like depends a lot on clothing, activity, length of time out, and age. Even two people dressed the same and in the same wind would not be likely to agree, if asked to estimate how cold it feels. Notwithstanding the variation in individual opinions, the figure given is one computed precisely from a formula that came from physics measurements, not measurements on people.

After some searching for the origin of the computation, I came upon a discussion of wind chill by David Phillips, of the Canadian Climate Center[1] in the *Canadian Geographic* magazine. In correspondence he provided me with a number of papers and references.

The formula generally accepted and used by weathercasters is an empirical one, based on a set of measurements of more than four decades ago, on the rate at which heat is lost by a bottle of water exposed to winds of various speeds and temperatures. That the measurements have stood the test of time, with no more than "fine tuning" adjustments, is not surprising, since they were simple enough to be verified by anyone possessing a thermometer and a wind speed meter. The part of the matter that can be argued is how the physics measurements apply to persons out in the wind, in various activities and with various kinds of protection. As would be expected, numerous studies of that question have been made, all arriving at different recommendations. But for the daily wind chill report on the radio, the formula in its simplest form has survived, which amounts to saying that you and I lose heat as a bottle of water does—possibly a poor approximation!

Getting from the data to a formula for wind chill equivalent temperature goes in two steps. The first is to fit a formula to the measured heat loss by the water bottle. That was done by the original experimenters. They gave it as:

$$H = [10.45 + 10(V^{1/2}) - V] (T_s - T),$$

where H is the rate of heat loss in kilocalories per hour per m² of exposed surface, V is the wind speed in meters/second, T is the air temperature in degrees Celsius, and T_s is the temperature of the exposed surface.

The second step is to get from that formula to the wind chill equivalent temperature, T_e. The reasoning goes as

```
JLIST

10   PRINT "ENTER THE TEMP, DEG. C";
20   INPUT T: PRINT
30   PRINT "W-SPEED KM/H","W-CHILL DEG C"
40   PRINT
50   FOR X = 1 TO 6
60   V = V + 10
70   A = .478 + .237 * SQR (V) - .0125 * V
80   WC = 33 - (33 - T) * A
90   PRINT "   "V,"   " INT (WC)
100  NEXT X

JRUN
ENTER THE TEMP, DEG. C?-10

W-SPEED KM/H      W-CHILL DEG C

    10                -15
    20                -23
    30                -28
    40                -31
    50                -33
    60                -35


JLIST

30   PRINT "W-SPEED MPH","W-CHILL DEG F"
60 V = V + 5
70 A = .478 + .301 *  SQR (V) - .02 * V
80 WC = 91.4 - (91.4 - T) * A
```

Fig. 1. Computer program, in Basic, by which a column of wind chill effective temperature for any given air temperature can be printed out, such as the example shown for −10°C. Alternate lines that can be substituted, for use with units degrees Fahrenheit and miles/hour, are shown.

follows: H is found by solving the equation with the measured air temperature and wind speed, and with T_s set at 33°C (to represent skin temperature, about 4°C below internal body temperature). Then the equation is solved again, asking the question, "What would the air temperature have to be to give the same rate of heat loss, if the wind speed were reduced to walking speed, 1.79 m/s (4 mi/h)?" Reasons for choosing 33°C and walking speed instead of zero are not quite clear, but they are accepted as the convention.

The above steps are combined into one equation for the wind chill equivalent temperature,

$$T_e = 33 - [10.45 + 10(V^{1/2}) - V] (33 - T)/22,$$

and it can be programmed into your computer to run off the values for a column of air speeds, at a given air temperature. The program in Fig. 1, for Celsius degrees and kilometers/hour, can be changed to work in degrees Fahrenheit and miles/hour by substituting the lines given below it. The result should check with what you hear on the radio.

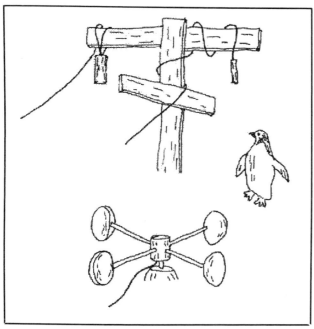

Fig. 2. A sketch, from the original photograph, of the water vial and the reference thermohm on a pole. Also a cup-type anemometer, the kind used in the experiment, is sketched.

Note that if you put in zero for the wind speed you get a nonsense answer (the result of using walking speed as the base in making the formula).

Interest in finding a wind chill "index," as it was called earlier, goes back at least to 1919.[2] An early one was based simply on the rate of cooling of a bare thermometer, through a few degrees downward from skin temperature, in the wind and out of it. Another did not even involve a rate of cooling—just the air temperature in number of degrees below zero C, multiplied by the wind speed in meters/second.

The experiments from which came the formula in common use today should be interesting to physicists. They were started in 1939 by Paul Siple, polar explorer and geographer, when stationed in Little America as a member of the U.S. Antarctic Service. During 1939–41

he, with his partner Charles Passel, made a long list of measurements on the rate of heat loss from a vial of water, under a range of conditions of wind speed and air temperature. There was no lack of opportunity: In no place on Earth is there more wind chill than in Antarctica in winter. The full data and the empirical formula for the rate of heat loss that is used today were published in 1945.[3]

As to the experiment, a cylindrical plastic vial containing 250 g of water was suspended from a pole outside the "science building." A sketch drawn from the photograph in the 1945 paper is shown in Fig. 2. The rate of heat loss was determined from the length of time required for the water to change from liquid to solid at the freezing temperature, in other words from the latent heat of freezing. The temperature was measured by the change in resistance of a Leeds and Northrup "thermohm," read by means of a Wheatstone bridge and galvanometer—inside the warm science building, of course. (A thermohm is simply a conductor that has a large temperature coefficient of resistance.) One thermohm was immersed in the water in the vial (left end of the cross-arm) and another (right end) was outside in the wind. The wind speed was measured at short intervals by a cup-type anemometer (lower sketch in Fig. 2), equipped with electrical contacts which caused a lamp bulb inside the building to flash at each revolution. The flashes in a given time were counted. More than 80 runs of data were taken, each with the water changing from liquid to ice.

It's a long jump from the freezing of water in the Antarctic night to the shivering of a person bundled in a coat on a city street. Of particular question is the use in the formula of 33°C for the temperature of the exposed surface, supposedly normal skin temperature. That suggests that the result should apply to a person with no clothes. And on that point there has been no lack of writing and comment. But, in our modern way of life and with our high efficiency clothing, the formula may in fact be more nearly correct for us than it was for the arctic explorer for whom it was intended. In walking the few blocks from car to office on a wintry day, we judge the discomfort mainly by the cold wind on our faces and ears. So the 33°C may not be far wrong after all. And when it's cold we walk fast, so the 1.79 m/s may be right as well.

References
1. 4905 Dufferin St., Downsview, Ont., M3H 5T4, Canada.
2. Leonard Hill, Med. Res. Council Repts. 32 and 52, London 1919, 1920.
3. Paul A. Siple and Charles F. Passel, Proc. Philosoph. Soc. of Amer. **89**, 177–199 (1945).

Doppler Radar: The Speed of the Air in a Tornado

Have you ever thought about how you might measure the speed of the air in a tornado funnel? If you were to set up an instrument and wait for a tornado to hit it, you might wait a lifetime. Surprisingly though, one investigating crew succeeded in that—almost. The tornado came close enough to record 30 m/s (67 miles/hr). Wind tunnel tests performed later showed that their instrument would not have survived had the encounter been closer. In years past, many attempts have been made to take high-speed movies of the debris flying around the outer wall of the funnel. Some have succeeded. The results have shown speeds of a little over 100 m/s (222 miles/hr). Because they are not direct measurements of the wind speed, the question remains as to whether the wind goes even faster than the debris.

In more recent times, a better method has been used: the Doppler shift in radio waves scattered back toward the source by the flying debris and/or rain in the funnel. This method has the great advantage that it can be used on tornados at distances of many kilometers; therefore, it is possible to chase one after it is sighted, and to set up and operate the equipment at a safe distance. The precision in measuring the speed is great, but as before, the measurement is not directly on the wind. And if the tornado is very far off, the tip of the funnel, where the speed may be the greatest, is not likely to be visible. The Doppler measurements agree generally with the photographic ones in showing speeds up to about 100 m/s. That measurement applies to the horizontal component of velocity; the vertical component, thought to be about half as much, is not shown by the Doppler method as used horizontally.

To understand the Doppler measurements, we have to rethink the Doppler shift from the way it is presented in the physics text. There the shift in frequency is worked out for a moving source and stationary observer and vice versa; typical examples are the sound heard from the whistle on a passing train, and the sound from the bell at a crossing as heard by a person riding on the train. For the tornado observations, the radio transmitter and receiver are not moving with respect to each other; they are stationary and next to each other. The radio wave gets from transmitter to receiver via debris being carried around in the funnel, as shown for one such object in Fig. 1A. The wave is scattered in all directions, and only a little of it comes back to the receiver.[1] The path between transmitter and receiver is said to be *folded*. If the scatterer is moving—as of course it is—the total folded path is increasing or decreasing, so the received wave is Doppler-shifted in frequency.

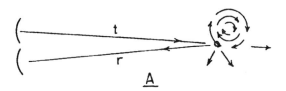

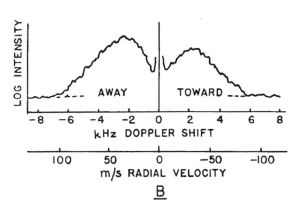

Fig. 1A. The path (t) is that of a radio wave from the parabolic transmitter dish to a piece of flying debris in the funnel of a tornado. The radiation is scattered in all directions, including the path (r) back to the receiving dish. The object shown is only one of many in the funnel, all of which scatter. Fig. 1B. A plot typifying the Doppler shift of the returned signal. Because the scattered radiation is received from a large number of flying objects, having a range of radial velocities, the total returned signal is a spectrum. The dashed lines indicate the background noise level. The lower abscissa scale gives the radial velocity corresponding to the Doppler shift on the scale above it, for a radio wavelength of 3 cm. The greatest radial velocity shown is about 100 m/s.

Having the transmitter and receiver side by side and fixed gives an enormous advantage in precision over the "nonfolded" case, such as the example of the train whistle. The reason is that the electronic circuitry can compare the (Doppler-shifted) frequency of the wave being received with the frequency of the wave being transmitted, and obtain the difference, or the *beat* frequency. The beat is the Doppler shift—the absolute shift, not the relative, or percentage shift.

There is a simple way to relate the radial velocity (the component of velocity toward or away) of the scattering object to the beat frequency.[2] When the radial distance to the scatterer increases by half a wavelength, the folded path length increases by a full wavelength. That increase makes the returned wave change from being in phase with the outgoing wave to out of phase and back to in phase, which translates into one cycle of the beat. So the radial velocity of the scatterer is a half-wavelength times the beat frequency.

An example will show the advantage of having the transmitter and receiver together so that a beat can be obtained. Suppose the transmitted (radio) signal is 3-cm wavelength, which is a frequency of 10 gHz (10^{10} Hz), and that the radial velocity of the scatterer is 10 m/s. The beat frequency will be 670 Hz. That easily can be measured to ± 10 Hz, making the radial velocity known to ± 15 cm/s—a high accuracy, indeed.

Although experiments on tornados using the Doppler effect go back nearly 40 years, there have been recent improvements toward greater mobility for the equipment and greater precision. Two of the workers in the forefront of those efforts are Howard Bluestein of the University of Oklahoma and Wesley Unruh of the Los Alamos National Laboratory. I'm indebted to them for information on their latest experiments.[3] Their transmitter sends a continuous wave (CW) of 3-cm wavelength. The antenna is at the focus of a 50-cm parabolic dish which directs the radiation in a beam of $5°$ width (at 3 dB). The receiving antenna is in an identical dish, pointed in the same direction. The beat signal, from debris moving with typical tornado velocity, is in the audio range, so it is recorded on a simple tape recorder for later analysis. All the entire equipment is battery-powered, and the pair of dishes is mounted on a tripod so as to be steerable. When chasing a tornado, the outfit can be set up on the ground and put into operation in one minute.

The received signal is not a single Doppler-shifted frequency: The transmitted wave is scattered by a large number of bits of flying debris, all having different radial velocities. Therefore, what is received is a *spectrum*. The content of the tape gives a spectrum such as sketched in Fig. 1B. The two sides are purposely not shown as symmetrical; they may not be for two reasons. The beam of radiation may not be centered on the tornado funnel, and the tornado as a whole may be moving toward or away.

Some general remarks about the kind of system just described may be of interest. Because the transmitter is on continuously, it is commonly called CW (continuous wave) radar. It is also called CW Doppler radar. The "radar" part is something of a misnomer. It is the acronym for **ra**dio **d**etection **an**d **r**anging. The CW system does not distinguish range: Moving objects at all ranges within the beam give beat signals.[4] That does not cause a problem in its use for tornadoes, because nothing else within the beam is moving with comparable velocity.

The simple CW version has had many applications. Early in World War II, work was done on it with the object of detecting approaching enemy airplanes. I witnessed a test at the Boston airport. The radiation was of 20-cm wavelength, directed by a parabolic dish. The receiver was near the transmitter, also using a dish. An airplane approaching at 100 m/s would give a beat note of 1 kHz, heard on a loudspeaker. The system worked: Even a pigeon was detected (with lower beat). But within a short time the method lost out to pulsed radar, which gave the important information of range. A CW Doppler system using not radio, but ultra high frequency sound, is used in a number of ways—for example, an intrusion alarm. Also in a most interesting way, such a system can measure the rate of blood flow in a blood vessel by means of an ultrasound transmitter and receiver outside the skin.

References
1. The intensity of the scattered radiation falls off rapidly as the greatest dimension of the scattering object falls below a half wavelength, which in this case would be 1.5 cm. But it does not go to zero: Even hail and rain give detectable scattering at the wavelength.
2. Where the velocity of the transmitter or receiver is not greatly different from the velocity of propagation of the wave, as may be the case for sound, the textbook Doppler formula should be used. In the case of radio waves, the speed of propagation is so high compared to the speed of the scatterer that the simpler calculation is accurate enough for any purpose.
3. H.B. Bluestein, Univ. of Oklahoma, and W.P. Unruh, Los Alamos National Laboratory, "Use of a portable CW Doppler radar to estimate windspeeds in severe thunderstorms," 15th Conf. on Severe Local Storms, Feb. 22–26, 1988, Baltimore, Maryland.
4. In applications in which a measurement of range as well as radial velocity is needed, the Doppler feature is added to the conventional pulsed radar. In brief, the pulse, which gives the range, is lengthened (with the loss of some range sharpness) enough so that the Doppler shift can be measured within it, to get the radial velocity. When your TV weatherperson says the prediction comes from "Doppler radar," it probably is pulsed Doppler.

On Electric Shocks and Spinning Eggs

I received a letter from Antoine du Bourg of the Pingry School[1] saying he used what I have often called reverse engineering on a gadget and got an unexpected result. Part of the way into the process, a complex of springs and small parts flew apart so he could go neither forward nor backward. The object is called Shock

Sentry, or GFCI (ground fault circuit interrupter),[2] and is essentially an automatic circuit breaker. If you start to get a shock or if some other small leak to ground occurs, it disconnects "In less time than that likely to harm a person coming in contact with the shock," the description says.

The key Tony du Bourg found as to how the device works was a ring of ferromagnetic material through the center of which ran both of the conductors on their way to the load. There was a winding of wire on the ring in the form of a toroid. Evidently, it builds on a standard method of measuring the alternating current in a circuit, as in Fig. 1A. In that method one of the current-carrying wires goes through the ring so that it and the return part of the circuit form a single turn that links the ring. A secondary coil of many turns wound on the ring gets an induced voltage. It is a simple transformer. The output, read on an ac voltmeter, is proportional to the current in the single turn. For practical reasons the ring is split and hinged so it can quickly be snapped around any wire.

The difference in the Shock Sentry, as Tony found, is that both of the wires connecting the load go through the ring (Fig. 1B). So, if all is well, the currents in the two wires are the same but in an opposite sense, and no emf is produced in the secondary coil. Another way to see why no emf is produced is to note that the current circuit does not link the ring. But if there is leakage (possibly through a human body) so that some of the current (dashed line a) returns by a route other than through the ring, there is no longer complete cancellation. In other words, some of the current does link the ring. So a voltage appears in the secondary. That acts to trip the circuit breaker.[3] The literature says that a five mA difference in current will trip. For such sensitivity, we assume that amplification is required so that is indicated in the figure. There is, of course, a reset button, and there is also a button that will introduce a small leakage to test the tripping.

It should be noted that for current to bypass the ring via the ground and create the unbalance, there must be a path to ground on each side of the ring. If there were no path to ground to the left of the ring (in the figure), then in principle even a person with bare feet on a wet floor carelessly touching a wire to the right of the ring would get no shock. But, in fact, that lucky situation would seldom arise, because in the standard house circuit one side of the 110 V supply line, called the neutral, is grounded. So, typically, for the careless person who touches the hot wire, there will be a completed circuit that links the ring so as to make the Shock Sentry trip.

Let me end this piece by suggesting an experiment—one that doesn't involve electric shocks. Most physics teachers are aware of the contention that an egg will stand on end at the moment of the equinox

(when the sun is in the Earth's equatorial plane). Often pictures appear to prove it. The latest "proof" I have seen appeared in the *Harvard Magazine*,[4] where four eggs are shown standing on end.[5] But, no fair! They are on rough brick paving; two are actually standing in the grooves where the mortar is. The article cites a record of 14 eggs standing at once (no such entry in Guinness).

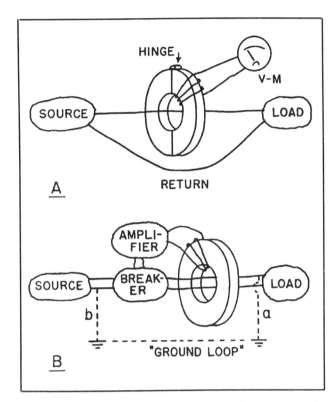

Fig. 1A. A well-known method of sensing the ac current in a line: closing a ferromagnetic ring around one of the conductors, making it the equivalent of a transformer core with a one-turn primary. Emf is induced in a secondary wound on the ring and indicated by a meter.

Fig. 1B. A variant of Fig. 1A, in which both wires going to the load go through the ring. The currents are equal and opposite at all instants so no emf is produced—unless there is a leakage path that allows some current to circumvent the ring. Such a leakage path might be a person standing on the ground and touching the "hot" wire.

The above was for fun; there is a phenomenon about eggs on end that poses a sensible question of physics. It is the very useful test of whether an egg is fresh or hard-boiled. It is simply to give the egg a twirl with the fingers, and if hard-boiled, it will spin on end like a top. If fresh, it will not only fall over but will do so within a fraction of a revolution. I have tried to estimate the angle through which the finger-twirl acts and come up with about 90° at most. That suggests that not much more than the shell gets the twirl; most of the (semiliquid) content not following. So, of course, it falls over—it is not really spinning.

Here is what I propose that you try. Make some device, maybe an adaptor for an electric drill, by which the fresh egg can be put into a spin throughout, on end of course, before being released. Then, will it spin like a top or is there a further reason it will be unstable? If you do not want to risk a mess, a plastic Easter egg that comes apart in the middle filled with water might do. Space in this column awaits the result!

References
1. Martinsville Rd., Martinsville, NJ 08836.
2. Eagle Electric & Manufacturing, Co., 45-31 Court Square, Long Island City, NY 11101.
3. The standard method of tripping: the circuit breaker is held closed by a mechanical latch. Force to trip the latch is exerted by an electromagnet (solenoid).
4. March-April issue, 1987.
5. Photos by Barbara Cerva of Harvard's Biological Labs.

More Flying, Spinning Objects

As a result of my earlier discussions of the Frisbee and the Aerobee,[1,2] Stephen Weiss[3] sent me a kit of do-it-yourself flying rings he designed, called Flite Rings.[4] Stephen Weiss has been a life-long innovator of origami (folded paper) objects, especially those that fly.[5]

The kit contains the makings of five rings in the form of flat sheets of plastic-laminated cardboard, precut, and having dented lines for bending. The rings are polygons of various shapes and numbers of sides with a maximum diameter 33 cm. Assembly is easy—a few minutes for each. You throw it, spinning, in the horizontal or up-sloping plane the way you would throw a Frisbee.

Note, in Fig. 1, the downward slope of the surface both ways from the crease line. The "roof shape" constitutes a sort of air foil, that gives the ring lift in flight. The lift seems to be quite strong and that, combined with light weight, allows the ring to stay aloft at low speeds, to "hover" as the literature says.

Seeing the finished ring may call up a question of topology. How can a single flat piece of material be formed into the three-dimensional ring of Fig. 1 without further cutting or tucking or wrinkling? The answer may be suggested by considering a property of a surface that is a part of a true cone. Such a surface (shown in Fig. 2 as B) is made by closing the V-gap in the flat piece of material, shown as A. Then it may be turned inside-out (as you would a sock) on the dashed circle to make it re-entrant, shown as C. Since the surface s' becomes the mirror image of what it was at s', no wrinkling is required.

The Flite Rings are close to the above geometry, close enough to require little yielding or forcing of the material. The fact that they are polygonal instead of circular makes for a couple of practical advantages, as I found in assembling them and getting them to fly right. The first is that the bending required to make the roof shape is on several straight lines—far easier than trying to turn the cardboard cone of Fig. 1 inside-out along the dashed circle!

The second advantage of the straight creases is that, with a little more bending by the fingers, it is easy to make small changes in the slope of the outer part of the ring between one test flight and the next. The purpose is to change the relative lifts of the leading and trailing parts of the ring and, therefore, to move the center of lift forward or backward along the line of flight. For example, if the outer slope of the "roof" is made a little steeper, the center of lift is moved backward with respect to the direction of flight. Unless the center of lift is adjusted to be close to the center of gravity, the ring will not stay horizontal in flight but will tilt and go down on its edge. If it tilts, gyroscopic precession is the culprit, as analyzed in the earlier column on the Frisbee.[1]

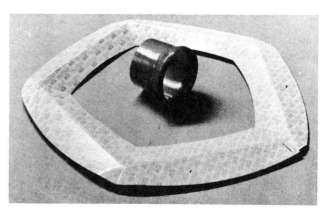

Fig. 1. One of the Flight Rings and the Zinger. Note that this Flight Ring is joined by a staple. The kit comes with a supply of wide double-stick, which makes a neater job.

So much for the Flight Ring. Look now at the tube in the same photograph, a newcomer to the toy stores, called a Zinger.[6] You throw it with the beaded end forward, spinning, as you would throw a football in a forward pass. It maintains its axis approximately horizontal and pointing approximately forward. The blurb on the package says it will go 100 yards. My far more modest results suggest that the 100-yard record might have been made by a major-league baseball pitcher.

The designer, Ronald Etheridge, wrote me that he got the idea from watching kids throw cardboard soft drink cups with the bottoms removed. Then he optimized the shape by numerous trials before having it molded in plastic.

A little further investigation indicates the flying tube may be one of the most frequently re-invented of toys. The Greeks may have had them. A recent catalog[7] lists one called the Toobee, made of sheet metal. An earlier one sold as a toy was called Skyro.[8] The tubes have been favorites of the paper-folders[9] since they are easily rolled up out of a single piece of paper. One version includes a fin, like the dorsal fin of fish, except it is set at such an angle that it maintains the spin. A feature common to all the versions is extra mass at the front rim, placing the center of mass forward of the center of the tube.

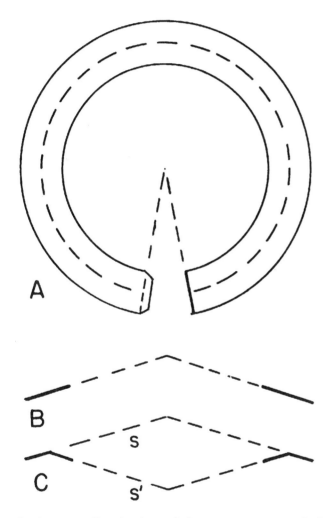

Fig. 2. How a flat ring is made into a re-entrant conical surface, the principle used (almost) for forming the Flight Ring.

The physics of the flying tubes involves, of course, gyroscopic precession. Attempts to analyze that have appeared in this journal.[8,10] The physics is complicated, but a few simple tests can be made just by throwing, which I have done.

The first test: to see if the forward center of mass is required for maintaining the orientation. The Zinger thrown rear end forward, with spin, tumbled. A tube without extra mass at either end (the middle part of a 7-UP can) also tumbled.

Second test: the Zinger thrown without spin. It maintained its axis on the direction of flight. But it behaved like an arrow; its axis followed the line of flight so that its "angle of attack" was zero, and it had no lift. It followed approximately a parabola, as would a stone.

Third test: the Zinger with spin. Two effects were evident. The gyroscopic action kept the axis from turning downward (at least in the early part of the flight) so it had a positive angle of attack and lift. When thrown with clockwise spin (as seen by the thrower), the tube, and therefore its path, turned to the right. Assuming that the turn represented gyroscopic precession, it is deduced (with the well-known angular momentum vectors) that the center of lift is forward of the center of gravity.

You too can have fun with tubes, and they can be made from almost anything. I loaded the front end of the tube from the 7-UP can with two diametrically opposite screws; it made a good flight. An interesting question: suppose the loading were adjusted to make the center of gravity and the center of pressure coincide? Would precession be eliminated? Maybe not, for the center of pressure probably shifts with speed and angle of attack.

References
1. *Phys. Teach.* **21**, 325 (1983).
2. *Phys. Teach.* **24**, 502 (1986).
3. 570 South Shore Dr., Miami Beach, FL 33141.
4. Kit by AG Industries, 3832 148th Ave., N.E., Redmond, WA 98052.
5. Stephen Weiss, *Wings and Things: Origami that Flies* (St. Martin's Press, New York, 1984).
6. The Zinger Co., 170 Gaines Ct., Athens, GA 30605-4144.
7. *Into the Wind*, 1988 catalog, 1408 Pearl St., Boulder, CO 80302.
8. Robert Liefield, *Phys. Teach.* (letter to the editor) **17**, 334 (1979).
9. Keith R. Laux, *The World's Greatest Paper Airplane and Toy Book* (Tab Books Inc., Blue Ridge Summit, PA, 1987).
10. "Questions Students Ask," *Phys. Teach.* **16**, 662 (1978).

What Does the Drinking Bird Know About Jet Lag?

What *does* the drinking bird know about jet lag? Maybe nothing. But the question is worth a closer look. Jet lag has interested me for a long time because of the way the behavior parallels that of a simple-systems class in physics. What tips the balance for me toward writing about it in this column is a statement heard or read several times recently[1] – that there is an east-west asymmetry in jet lag. Supposedly, it takes longer to recover after a trip to the east than after one to the west. Surprising, but that also comes out of the parallel with simple physical systems.

The physical system we start with is the drinking bird — the one from the toy store, not the one traveling in the jet. It is an example of a class of devices or systems known as relaxation oscillators. More examples are everywhere you look, from the dripping water faucet to the beating of your heart. Figure 1 shows a variety of examples, most involving liquids, chosen because the liquid ones seem to be the simplest kind to explain. They all differ in a basic way from the kind of oscillator commonly studied, such as the pendulum.[2] First see how they work, and later see if they have anything to do with jet lag.

To return to the drinking bird, I've re-sketched it (Fig. 1A) from a figure that appeared in this journal long ago.[3] The bird is a heat engine, deriving its power from a difference of temperature between the bulb (body) and the head. The bulb is at room temperature, but the head is cooler, due to evaporation of water from the felt on the outside of the head and beak. The cooling depends on the head being wet, which depends on the beak dipping into the water regularly. Due to the temperature difference, the pressure of the vapor in the bulb is greater than it is in the head, so liquid[4] is forced up the tube (upper diagram), shifting the center of gravity so the head goes down and the beak dips (lower diagram). In that position the lower end of the tube is open to the vapor, so the liquid can drain out of the tube, back into the bulb. The bird tips upright, and the cycle starts over.

The geyser (Fig. 1B) is another heat engine. Water from an underground spring slowly fills the hole which extends down into hot volcanic magma. The water at the bottom gets heated far above the normal boiling point of 100°C, being prevented from boiling by the pressure of the water column above it. Finally it becomes unstable; the first steam bubbles form at the bottom, forcing some water out at the top. That reduces the pressure at the bottom, because the column is now part steam. The reduced pressure lets the boiling accelerate, and the whole water column blows out. Then it starts over.

The next one (Fig. 1C) as yet untried, is suggested by the water box of the venerable flush toilet. A slack string

from the float connects to the hollow rubber ball that closes the outlet. A small spring is added in the line, to give it "snap action." A small stream of water enters continually, and it flushes periodically. (Pranksters note: application to a real toilet might take some experimentation.)

Another example from hydraulics is the ram (Fig. 1D), described some time ago in this column.[5] Water in the long inclined pipe accelerates, until it suddenly raises the vertical spindle and disk (a), stopping the escape of the water. (The spindle valve is shown in the up, or closed, position.) Due to its momentum, water pushes through the one-way valve (b), adding some water to the ballast bulb and thus to the elevated tank. When the momentum has been expended, the pressure is low enough so the valve opens (drops by gravity), and the cycle starts again.

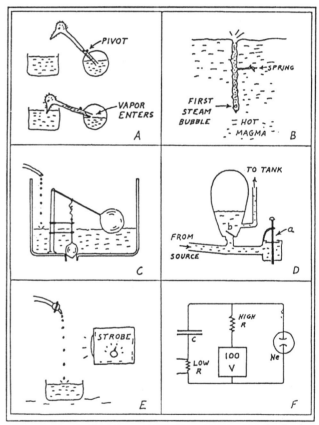

Fig. 1. Several drawings that represent simple systems in physics.

Next is the familiar water dripper, (Fig. 1E). Here surface tension is in control. It supports the drop as it grows, until its weight reaches a certain critical value. Then it breaks away, and a new drop starts to grow. Often the drops are made to "stand still" by illuminating the dripper with a strobe lamp.

Finally, the well-known neon flasher (Fig. 1F). A capacitor slowly charges until at about 60 V the neon lamp suddenly glows and conducts, discharging the capacitor down to about 40 V, at which point the neon glow extinguishes, and the capacitor starts to recharge. The next cycle begins.

If the oscillators described were to be used as clocks, their time-keeping accuracy would be found to be pretty poor. But they have the interesting property that they will lock into step with a slight nudge, applied with regular timing, from the outside. Here is how.

We may imagine that the bird is given a gentle tap on the head when he is nearly ready to dip on his own accord. If he is near to dipping, a very slight tap will suffice to trigger the dip. (The nearer he is, the more unstable, and the more easily triggered.) Suppose the tap is given each time he is nearly ready. Then he will no longer be "free running" but will be under the control of the taps. And if the taps are derived from a precisely regular source—like a quart crystal, the bird will be changed from a somewhat fuzzy timekeeper to a precise one. Because the tape terminates each cycle a little early, the frequency with which the taps come must be a little higher than the bird's free-running frequency. The importance of that will come out soon.

The other relaxation oscillators in Fig. 1 can be controlled and made into precise timekeepers in a similar way. But each requires a different kind of "tap": extra heat at the bottom of the geyser, maybe from a little electric heater turned on momentarily; a small voltage pulse to the neon flasher, applied across the "low R" in the figure, and so on.

Finally, we get to the jet lag. Every creature, as you know, has an internal "clock" that operates on a cycle, in most cases daily, called the circadian rhythm.[6] It involves the biochemical processes that control activity, sleep, etc. It is not simple like the relaxation oscillators of Fig. 1, but there are some remarkably parallel characteristics:

1. If the biological clock is allowed "free run" it is found to be a pretty inaccurate timekeeper. That observation can be made by isolating the creature from all outside influences—for example, by keeping it in a dark cave.

2. When not isolated, the clock "locks in" to signals from the outside, such as the coming of daylight. That keeps it in phase with the day-night cycle.

3. In tests on living creatures in isolation, it has been found, invariably, that the free-run frequency is a little lower than the day-night frequency. The free-run period for people is 25 to 26 hours. That checks with the requirement for lock-in as described for the drinking bird.

4. This is the parallel I find the most interesting: the asymmetry in recovery time for travel—east vs west. In a simple system like the drinking bird, if the phase of the controlling taps on the head is displaced, say a quarter of the cycle one way or the other (analogous to a person taking a quick trip through six time zones) the lock-in will be lost, leaving the bird to free run. The lock-in will be regained after some length of time or a number of cycles, because the two frequencies— taps and free run—are a little different. But, as you may see by making a series of marks on each of two pieces of paper and sliding them with respect to each other, the time for the "catch-up" will be different for equal displacements forward or backward (a jet trip east or west).

The little experiment above even agrees that more cycles have to go by for re-locking to occur if the displacement is forward (trip to the east) than if it is the opposite.

Better plan your next vacation to the west!

References
1. "Tips For Overcoming Jet Lag," a folder by Upjohn Co., 700 Portage Rd., Kalamazoo, MI 49001. A feature article in the *Portland Oregonian*, June 12, 1988. The TV program "Newton's Apple," July 16, 1988.
2. They do not retain their energy from cycle to cycle, as does the pendulum.
3. Kemp Bennet Kolb, *Phys. Teach.* **4**, 121 (1966).
4. Methylene chloride, boiling point 40.1°C at normal pressure.
5. *Phys. Teach.* **26**, 245 (1987).
6. Ritchie R. Ward, *The Living Clocks* (Mentor paperback, 1972); Kenneth J. Rose, *The Body in Time* (Wiley & Sons, 1988); and Arthur T. Winfree, *The Timing of Biological Clocks* (Scientific American Library, 1988).

How to Measure the Flow of Liquids

Some time back, Albert Bartlett[1] sent me literature on two versions of a device for measuring liquid flow. One, especially, struck me as clever and interesting because it can be used for acids or other corrosive materials. In this device, no electric wires or rotating shafts connect the liquid and the outside world, and the parts in contact with the liquid are made of nonreactive materials such as teflon. The other version is not adapted to use with reactive liquids, but it has the advantage of being able to measure a very wide range of flow rates in pipes of unlimited size. It is used mainly for water. I followed up by talking with the designer, Murray Feller, of the Wilgood Corp., the manufacturer.[2]

The first-mentioned version is built into a short tube or piece of pipe, which can be inserted in series in the pipe carrying the fluid whose rate of flow is to be measured. It is made in diameters from about 4 to 15 cm (1 ½ to 6 in). Figure 1A is a sketch of the working part. Inside a short tube, near each end, there is a set of stationary vanes, in screw shape, which give the liquid that is flowing through the tube a rotation, along with its forward motion. A double-ended cone, on the center axis of the tube, is supported by the blades. A pair of grooves, one around the center of the cone and the other around the inner wall of the pipe, form a raceway for a single plastic ball, which can move freely. The component of rotation of the fluid as it moves through makes the ball run around and around in the track, the number of rev/s being proportional to the forward velocity of the fluid. The obvious next question is: how is the number of rev/s measured, with no physical connections from inside to outside?

The small enclosure on top of the pipe in Fig. 1 contains a "pick-up" electrode, which is the antenna of a 10 MHz oscillator (a mini radio transmitter). The electrode is outside a barrier that separates it from the fluid. But when the ball passes near, just inside the barrier, the power load on the oscillator takes a momentary dip, or an increase, mainly because the plastic ball has a dielectric constant different from that of the fluid. The power blip is sensed, for example, through the change in the current from the power supply to the oscillator. As the ball goes around, the series of blips triggers circuitry in a microchip that counts the rev/s of the ball and computes a final readout in GPM (gallons per minute) if the unit is used in the United States, otherwise in ℓ/s.

The way friction of the ball moving in the track is minimized is interesting. The average density of the ball is made close to that of the fluids so that, in the words of the maker, it has nearly "neutral buoyancy." If the ball were to have exactly neutral buoyancy, in the fluid at rest, it would exert no force at all on the track. But that would be an unreal situation. In the moving fluid there is a force due to drag—directed downstream—and assuming that the buoyancy is not exactly neutral, there will be a force on the track from that cause, magnified due to the centripetal acceleration of the ball as it goes around its orbit. But the effects of both the drag and the centripetal acceleration approach zero as the flow velocity approaches zero, which gives the company valid reason to emphasize that the flow meter retains its accuracy down to velocities very close to zero.

We must look more closely at the matter of the centripetal acceleration, when the flow is fast and the ball is going around with high angular velocity. Will the radial force on the track be large—so large perhaps that the ball will have to roll? It might best be thought of in terms of a centrifuge, which in fact it is. If a solid object suspended in liquid in a centrifuge has neutral buoyancy to start with, it will stay neutral as the centrifuge rotates, no matter how fast. But if it is not quite neutral, the small net force that would tend to make it sink or float in the liquid at rest will be multiplied (in the radial direction) by the number of g produced by the rotation. That is how nearly neutral particles can be precipitated or be made to float to the surface.

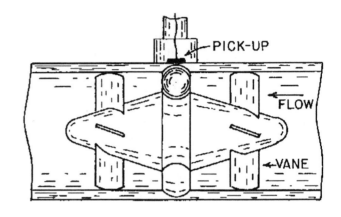

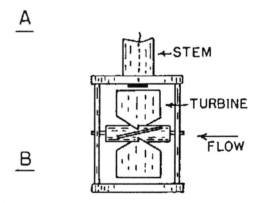

Fig. 1. A: The pipe shown cut away, to expose the double-ended cone, the stationary vanes, and the ball. The vanes are of screw shape. They give the passing fluid a rotation, and that drives the ball in orbit in its track. The pick-up electrode (antenna) senses the passages of the ball and sends pulses to a minicomputer, to be translated into rate of flow. B: A flow meter of open design, whose vanes rotate as a turbine. The pick-up senses the passage of the vanes and sends pulses to the minicomputer. With a sufficiently long stem, this kind of meter can be inserted into a pipe of any size.

The implication for the fluid meter is that yes, the ball may exert a force on either the outer or the inner track, but only in the amount by which it would tend to rise or sink in the fluid at rest, multiplied by the number of g from the rotation. Wondering how high that multiplying factor might be, I phoned Murray Feller for some data. For example, in a pipe on which the orbit radius is 3 cm, the ball may go as fast as 20 rev/s. That figures out to be about

50 g. Consequently, if the ball is not to exert a considerable force on the track, it must be adjusted to have very nearly neutral buoyancy.

The second version of the flow meter is somewhat simpler. As sketched in Fig. 1B, the working part is in the open, on the end of a probe stem. It is just inserted into whatever pipe carries the fluid so that the working part is near the center. As in the other version, the fluid passes a set of vanes that have screw shape but in this case the vanes rotate as a turbine. They are on a short shaft, with journal bearings at the ends, which are tungsten carbide pins running in sleeves of sapphire. Those two hard materials are used in order to minimize wear from abrasive particles that might be carried in the fluid. There, again, the average density of the moving part is made close to that of the fluid, to minimize friction. But in this case centripetal acceleration does not affect the friction in the bearings.

Sensing the motion of the moving part is done in a way similar to that in the meter first described. The passage of a vane of the turbine near a pick-up electrode makes a pulse; the number of pulses/s is translated into GPM or ℓ/s. The range of operation for the above as given by the manufacturer is "1 GPM in a 1 1/2-in pipe to 230,000 /gpm in a 72-in pipe." The interesting part is the lower limit, which shows that the device must be quite friction-free, due in part to the near-neutral buoyancy of the turbine. There is no upper limit on the pipe size or rate of flow if only the stem is made long enough. It could work in the Amazon River!

Fluid meters come in great variety. In an earlier column,[3] the kind used in a gasoline dispenser was described. It is of the kind called "volumetric." It is a little piston-and-cylinder engine, run by the passage of the gasoline through it. Each stroke of a piston measures exactly a quantity of gasoline, and the stroke at the same time sends an electrical pulse to a minicomputer that tells how much you have to pay. An interesting feature of the engine is that it has to have more than two cylinders, in order for it to be self-starting from any position in which it happened to stop. It has three cylinders 120° apart.

Murray Feller tells me that one variety of water meter used in homes is not unlike the second version we described. It has a rotating turbine, but the rotation is not sensed and computed electrically. Instead, a magnet is carried around on the turbine, and on every revolution it gives a pulse of force to a magnet outside the pipe. Those pulses of force act on a ratchet-and-gear train, which moves the readout pointers that tell how much your bill will be.

References
1. Albert Bartlett, College of Engr. & Appl. Sci., Univ. of Colorado, Boulder, CO 80309.
2. P.O. Box 1247, Dunnellon, FL 32630.
3. *Phys. Teach.* **25**, 168 (1987).

Mystery Glow-Ball: When Is a Battery Not a Battery?

Editor's Note: This article was contributed by Beverley A.P. Taylor and Barney E. Taylor, Miami University, Hamilton, OH 45011 as printed in *The Physics Teacher*, **27** (8), p. 630, ©1989 American Association of Physics Teachers.

Several months ago the following advertisement caught our attention:

Mystery Glow-Ball Lights Up in Your Hand! No batteries, no wires! Just hold ball in your palm and watch it begin to glow, then grow brighter! Actually uses energy from your own body for its power. Really mystifying; great for parties, magic tricks, etc. 1½-in diameter. Plastic.[1]

Inquiries of chemistry colleagues about light-producing chemical reactions that might be catalyzed by the heat from one's hand produced no answers or even speculations, so we decided to order one and dissect it.

The Glow-Ball package made no mention of the presence of or lack of batteries. It did include the following description, "The mysterious 'super energy ball' comes from the Milky Way. Hold it in your hand and it will brighten slowly and give you power and luck." (How many readers would laugh off "comes from the Milky Way" as nonsense without recognizing the truth in the statement?)

As shown in Fig. 1, the Glow-Ball resembles a Ping Pong ball with metal rivets about 15-mm apart in one side. The rivets look suspiciously like electrical contacts but could also serve to increase heat conduction. While experimenting with the Glow-Ball before opening it, we were intrigued by the existence of a noticeable delay before the glow begins. This could be due to the time required for sufficient heat to be transferred but was also reminiscent of a capacitor charging. By the time we opened the Glow-Ball, we were not very surprised to find two small batteries that provided energy to a small lamp. We decided to continue with this investigation and see just how much we could learn about the Glow-Ball.

Inside the Glow-Ball is a small, printed circuit board containing two 1.5-V cells, a three-transistor amplifier, and a small "grain of wheat" lamp. The circuitry is connected to the lower half of the Glow-Ball by the rivets in the bottom. A schematic of the circuitry is shown in Fig. 2. By inspection of the circuit, we conjectured that the following happens. When one's skin contacts the rivets, the base of transistor Q1 is pulled toward ground, causing Q1 to conduct. As soon as Q1 conducts, capacitor C begins charging. When the capacitor has charged to a sufficiently high level, transistor Q2 conducts, pulling the base of Q3 toward ground. This causes a current through the collector circuit of Q3, and the lamp begins to glow. Thus, the charging of the capacitor is indeed responsible for the time delay. When contact with the rivets is broken, the lamp does not immediately go out due to the effect of capacitor C. The stored charge in C will flow through R2 and R3 until the capacitor is discharged. As long as the voltage across C is sufficient to turn on transistor Q2, the lamp will remain on. As the capacitor voltage drops to about 0.7 V (the base-emitter voltage for a silicon transistor), the lamp will gradually dim as Q3 is slowly turned off.

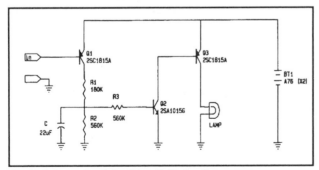

Fig. 1. The mysterious Glow-Ball.

Fig. 2. Schematic of the Glow-Ball circuit.

The supposed behavior just described was experimentally verified by connecting a microcomputer with a four-channel data acquisition system to four different points on the Glow-Ball, with ground as the reference. The first channel was connected to the collector of Q1. The collector was chosen as indicative of changes in the input because, due to the high resistance of the skin, the current at the rivets necessary to cause the lamp to glow must be extremely small. The second channel was connected to the high side of the capacitor, while the third channel was connected to the base of transistor Q3. The fourth channel was connected to the high end of the lamp. Data were typically acquired at 20 points/s for 15 s.

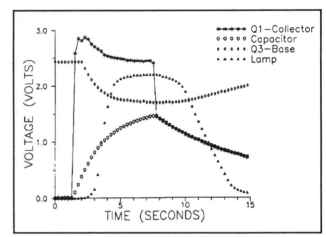

Fig. 3. Voltage-vs-time graphs at four points in the circuit as the contacts are "touched," then released.

Figure 3 is a subset of one run. In the interest of clarity, every fifth point is indicated. The rivets are "touched" at about one s into the run. The collector of Q1, initially at ground potential, rapidly rises to about 2.8 V. (It levels down to 2.4 V as the lamp comes on, presumably due to the load of the lamp on the batteries.) With the collector of Q1 at 2.8 to 2.4 V, the capacitor charges through the 180 K resistor, reaching about 1.4 V after 7 s, as seen in the plot. At that time, 7 s into the run, the rivets are released so the charging current to C ceases. The collector of Q1 quickly falls to the voltage of C, and together they fall in voltage as C then discharges through the two 560 K resistors in parallel — again seen in the plot.

The remaining two transistors, Q2 and Q3, receive a very small current from the capacitor and turn it into a high current for lighting the lamp. As can be seen in the plot, the rising voltage of the capacitor (and base of Q2) makes the collector of Q2 and base of Q3 go downward, and therefore charge flows through the lamp. At 7 s, when the voltage of the capacitor starts downward, the whole process reverses — also clear in the plot.

There is one interesting anomaly about the circuit. The lamp will not light if the touch contacts are shorted with a conductor. We believe that the emitter base junction of Q1 effectively behaves as a forward-biased diode, thereby shorting the batteries to ground. Currents of approximately 150 mA were observed by touching the probes of a digital multimeter in current mode to the contacts. Currents of this level represent extremely heavy loading of the small cells within the Glow-Ball. Tests were made with various resistances connected between the contacts. The lamp would light with as little as 10 Ω connected to the rivets; however, the current was still excessive—about 85 mA. Surprisingly, the Glow-Ball would light if the digital multimeter (DMM) in voltage mode was connected to the two rivet contacts. The stated internal resistance of the DMM is 20 MΩ. (Typical skin resistance over one cm is on the order of a megohm.) This piqued our curiosity as to how large a resistance could be connected and still cause the Glow-Ball to glow. A number of high-value resistors were tried across the rivets. The ball would glow (but not at full brightness) with a resistance of 75 MΩ. No glow could be observed using 90 MΩ. This simple circuit will function with input resistances of 10 Ω to 75 MΩ, which corresponds to a range of almost 7 orders of magnitude.

In order to convince ourselves that our understanding of the circuit was complete, a similar circuit using a light-emitting diode as a lamp and a 5-V power supply was constructed. Since the Japanese transistors in the Glow-Ball were not readily available, generic PNP and NPN transistors from Radio Shack "blister packs" were used. The circuit worked much better with some of the transistors than others. Since the transistors are used to amplify the small current through the skin into a current sufficiently large to cause the lamp to glow, they need to have a large beta. Beta is the ratio of currents in the collector and base and has approximately the same value as H_{fe}, which is often a more readily available parameter.

Thus, the Glow-Ball does indeed use batteries. Being generous, one could assume the advertising copywriter meant no batteries to be purchased and inserted, as required with many toys. However, we can find no excuse for "uses energy from your own body."[1]

Reading advertising critically and questioning how things work are habits to be encouraged. They are everyday-life survival skills and often provide a springboard for an excellent physics lesson.

Reference
1. Spencer Gifts Catalog, Jan., 1989, (Spencer Direct Marketing Company, Spencer Building, Atlantic City, NJ 08411, 1988) p. 15. Mystery Glow-Ball is $3.99.

The Noises Your Muscles Make

At this university, there is a quarterly publication called *Research News* that looks at interesting projects on campus. A recent issue[1] had an item on muscle sounds being studied by Dr. Daniel Barry and his associates in the University Medical Center. Sounds from muscles? I had never heard my muscles! So I went to the lab where the experiments are being done. I quickly became a believer when Dr. Barry put a microphone to my forearm. Then back at home, having found out what to look for, I succeeded right away in putting the signals from my muscles onto an oscilloscope. For a preliminary idea, look at Fig. 1, but wait a little to know how it was done.

The fact that was new to me, which led to some pleasant hours of experiment, is that muscle contraction is not steady, but is modulated at around 20 Hz, more or less. That means that if you clench your fist, there is vibration at that frequency at the surface of your arm. It can be sensed by a microphone or a phonograph pickup, amplified, and either heard from a loudspeaker or displayed on an oscilloscope. The vibration is called sound because it is in the frequency range of sound, not because it is strong enough to be heard ordinarily when you lift something. If you were to hear it, amplified, you would call it a rumble, rather than anything approximating a pure tone.

The reason the sound that is picked up is not a simple sine wave is interesting. A muscle, as we speak of it, is made up, at the next level down, of a number of *motor units,* each made up of a large number of muscle fibrils. The number of motor units may be large (e.g., in the biceps) or small (in the muscle actuating the eyelid). Nerves stimulate the motor units to contract. When stimulated, a motor unit contracts in a short train (five or so) of pulses. Then it can be restimulated. The frequency of the pulses in the train is the frequency we hear as the muscle sounds. When a muscle is working, say when lifting something, the many motor units are *not* stimulated and restimulated in unison, but in what seems like random phases. That is a reason the "sound" from many of them together has the character of narrow band noise—or a rumble instead of a tone. It is quite evident in Fig. 1.

I asked Dan Barry how the muscles manage to give a graded response, because it might seem from the above that the force exerted by a motor unit is an all-or-none business. He said that graded response is accomplished by at least three effects, in various degrees and combinations, according to the demand. Not all of the units work at a given time: many, or fewer, are activated as the need changes. A given unit may be stimulated more or less often. The frequency in the trains of pulses (the typically ≈ 20 Hz) may increase or decrease.

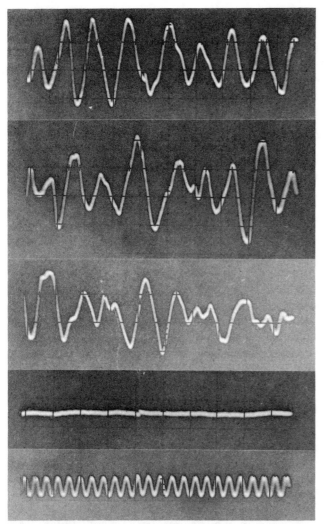

Fig. 1. Oscilloscope traces of muscle "sounds" —actually vibrations picked up at the skin by direct contact to a transducer. The top three were made with the transducer resting on the forearm and the fist clenched. Next below is a trace made with the transducer resting on foam rubber to show that extraneous electrical and acoustical pickup are insignificant. The final trace is of 60-Hz line frequency, for frequency calibration. (The breaks in the traces are due to a coordinate grid on plastic in front of the CR tube.)

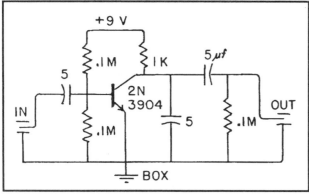

Fig. 2. The one-transistor preamplifier.

As it turns out, connecting up a circuit to demonstrate the muscle vibrations is simple if high fidelity, as would be wanted for research purposes, is not required. First, it is best to pick up the vibration by direct contact. A phonograph pickup head might be good if one is available. The transducer I used was a 1½-in (3.8-cm) cone-type speaker[2] from the junk drawer, probably originally from a $5 transistor radio; 2-in ones are listed by Radio Shack for $1.99. (A "tweeter" would *not* be suitable; its sensitivity at 20 Hz would be very low.) A slice from a cork (8 mm in diameter) is glued to the center of the cone of the speaker, projecting a little beyond the plane of the rim of the cone so that when the speaker rests upside down on the forearm most of the force is on the cork. It proves to be a good transducer, producing about 1 mV when resting on a tensed muscle.

The available oscilloscope has a maximum sensitivity of 10 mV/cm of Y deflection; therefore a preamplifier is required. A simple one, made with one general-purpose transistor, has ample gain: 30 at 20 Hz. Two features minimize stray electrical interference and pickup of sounds from the air. The band pass is restricted: it is down to half at 5 Hz and 60 Hz. The preamp circuit board, with a 9-V battery, is enclosed in a grounded metal box. As can be seen from the trace in Fig. 1, made with the transducer resting on a piece of foam rubber, there is very little stray pickup of any kind. The circuit of the preamp is shown in Fig. 2.

To get the traces in Fig. 1, the forearm was placed horizontally on the table, with the little speaker resting on it upside down so that the weight of the speaker was on the cork; it was not pressed down. The muscles were tensed by making a tight fist. It was necessary to search around, resting the speaker at different places, to find a muscle that gave good pickup. The single sweep of the scope occurred at 4-s intervals, giving time to open the camera and tighten the fist, in anticipation.

An approximate value for the frequency can be obtained from the traces of Fig. 1 by measuring peak-to-peak intervals, where the oscillations are fairly regular, especially in the top trace. The bottom trace, 60-Hz line frequency, serves for calibration. The result is 22 Hz.

Daniel Barry told me that muscle sounds have been known since the seventeenth century. In those times they were perceived by placing the thumbs against the ears and tensing the arm muscles. (You can check that: you will hear a low rumble, which increases as you tighten the arm muscles.) From the seventeenth century on, not much more was learned until modern electronic recording techniques came along. Now that the "signatures" of muscle action can be obtained easily, there is a field of application in the diagnosis of diseases and other abnormalities in muscles, as well as in the scientific study of muscle action. Leaving aside literature in the medical journals, the best source of information on the applications, for the layman, is an article in *Scientific American* by Gerald Oster.[3] It tells everything except how to do it yourself, which now you know anyway, from the preceding discussion.

References

1. *Research News,* January–March 1989. Published by the Division of Research Development and Administration, University of Michigan, Ann Arbor, MI 48109-1248.
2. Small speakers are, in fact, often made to double as microphones, e.g., in walkie-talkies and intercoms. Since they are of the moving coil type, the emf they generate is proportional to the velocity of the coil, and therefore, for a sinusoidal motion of a given amplitude, the voltage is proportional to the frequency. In most audio applications, that is compensated for in the circuit. In the use described here —just demonstration—compensation is unnecessary. Since the muscle signals are within a rather narrow frequency band, the slope of the response is of little matter.
3. *Sci. Am.* **250**, 108 (1984).

Sampling the Weather in the Upper Atmosphere by Balloon

Sending up balloons to get the profiles of temperature, pressure, relative humidity, and wind velocity in the atmosphere is a big, worldwide operation. In the United States there are on average two release stations per state, and worldwide the total is about 800. To make the data compatible, the times of release of balloons are the same for all participating countries: daily at 600 and 1800 hours GMT. For combination with its own data, the U.S. National Weather Service receives, within two hours after the standard launch times, the data from the USSR, Japan, China, India, and other nations. These data go first to a central computer in Maryland. From there, pertinent regional data are sent to the Weather Service stations throughout the United States for release via the local newspapers, TV, and radio.

Bob Snyder, manager of the Ann Arbor station of the National Weather Service[1] gave me this information and went on to tell more about the balloons and their flights. The rubber balloons are inflated with helium or hydrogen to about 2 m in diameter. They expand as they go up and burst at a height of about 30 km. The payload—the package containing the measuring instruments and the radio transmitter that send back the information—falls to earth on a small parachute, landing perhaps as far as 200 km from the starting point. A message on it asks the finder to give it to any mail carrier, no postage or wrapping required. When the flights are over land, 85 to 90 percent of the payloads are returned; they are then reconditioned for about 20 percent of the new cost.

On a windy day, the launching can be exciting. The balloon is inflated in a two-story shed that has a large door. The instrument package is attached to the balloon by a cord as much as 15-m long. If the wind is strong, the person doing the launching may have to run along behind (Fig. 1A), carrying the package until the balloon is high enough to keep the package from dragging over the ground.

A description of the sensing, transmission, and processing of the information would fill a book. The part most likely to interest physicists is how the temperature, pressure (which translates into altitude), relative humidity, and wind velocity are measured continuously as the balloon ascends. The instrument package that does all that and

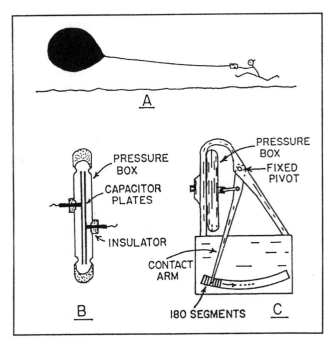

Fig. 1. A. How the instrument package sometimes has to be kept from dragging. B. Variable capacitor, the separation of whose plates is controlled by the atmospheric pressure outside the flexible metal box. C. An instrument like an aneroid barometer, having an arm that moves over 180 fixed contact segments, causing temperature and humidity information to be transmitted alternately. A count of the number of alternations gives the position of the arm, which translates into altitude.

transmits the data to earth is remarkably compact and light—about a quarter of a kg. All solid-state circuitry, of course. I owe thanks to two makers of the packages[2,3] for technical descriptions.

The sensors (except for those measuring wind velocity) work in one of two ways, depending on the company that makes them: by change in capacitance[2] or by change in resistance.[3] Let's consider the capacitance kind first. The temperature sensor is probably the simplest part. We are familiar with the disc capacitors, smaller than a dime, that we use in electronic circuitry. They are made with a ceramic dielectric, barium titanate, for example. The ordinary ones have quite a high variation of capacitance with temperature. Ceramics can be found that are even "worse"; that is, which have higher variation. The higher the better for this purpose. During a balloon flight, the temperature is sensed by such a capacitor.

The relative humidity measurement is also simple. There are many plastic polymers, or gel-like substances that take up water from the air (with some swelling) in proportion to the relative humidity. Water has an exceptionally high dielectric constant. So if a capacitor is made with such a gel between the plates, its capacitance will go up as the humidity goes up. Of course, the plates must be perforated to expose the gel to the air.

Finally, barometric pressure. That's the only measurement that requires a moving part. It starts with a circular, sealed metal box with flexible ends, as in the familiar aneroid barometer (Fig. 1B). A pair of capacitor plates is mounted inside the box so that as the box expands or contracts with change of atmospheric pressure the plates move closer together or farther apart, changing the capacitance.

How capacitance or resistance changes get translated into modulation or pulse trains, transmitted to the ground station by radio, and decoded by computer is state-of-the-art electronics. We have to leave it at that, except for one point: how several trains of data can be transmitted throughout the ascent of the balloon by a single radio. It's done by *multiplexing*, a standard way in which samples of the different data are transmitted in rotation, round and round.

Now we switch to the system that works on changes in resistance and start with the sensing of temperature. There are some materials—oxides of certain metals—that have very steep changes of resistivity with temperature. They are the *thermistors*. Once calibrated, they are very stable. They make a simple transducer for temperature in the balloon package.

Sensing relative humidity is more tricky. It calls for just what we have already talk about: a plastic, a polymer that takes up moisture from the air and swells. For this purpose, a strip of the plastic is coated very thinly with carbon particles. When it swells, the particles press against one another less, causing an increase in the resistance of the

carbon layer. One is reminded of the carbon granule telephone transmitters of not so long ago; some are still around, in fact.

Pressure is sensed by the kind of aneroid barometer already described. But getting the result to earth is done in a novel way. One might say for free. The pressure sensor does the multiplexing for the other two sensors and, in so doing, reveals its own value! If that sounds hard to believe, see Fig. 1C and read on.

The pointer, linked to the pressure box moves over an array of 180 contacts. Alternate contacts conduct to the radio transmitter the information from (say) the temperature sensor, and the other alternate ones conduct that from the humidity sensor. As the balloon goes up in altitude, the pointer moves from one contact to the next. The count of the number of alternations tells the position of the pointer, which translates into altitude. Thus the ground station gets all three kinds of information.

Only the tracking of the balloon to log the wind speed remains. Tracking by radar has been the workhorse method. But a more recent way is to make use of the worldwide Loran-C, Omega, and VLF radio location systems. In these public service systems, synchronized radio signals are transmitted from multiple locations at all times so that the location of a receiving station can be found from the relative times of arrival of signals, by triangulation. Users are mainly ships and airplanes. Tracking the balloon requires an extra step. The signals are received at the balloon package and retransmitted (transponded) to the ground station. Comparison of the signals received directly at the ground station with those transponded from the balloon allows for computation of successive positions of the balloon and thereby the wind velocity.

Weather balloons get pretty big before they burst—so if you watch the sky at the right time (1800 hours GMT is 1 p.m. EST) you may see one.

References

1. Bob Snyder, Federal Building, Ann Arbor, MI 48104.
2. Vaisala Inc., 2 Tower Office Park, Woburn, MA 01801 (world's largest supplier of atmosphere-sounding equipment). Information supplied by John T. Crocket.
3. VIS Manufacturing Co., 335 E Price St., Philadelphia, PA 19144-5782. Information supplied by A. Scott Bell.

The Stirling Engine—
173 Years Old and Running

My interest in the "hot-air engine" was rekindled when, a decade or more ago, I received an ad for a small, beautifully finished, operational model. Needless to say, I

sent for it[1] (Fig. 1). At that time I thought of a hot-air engine as one of the many engines that were invented and then left along the road to the internal-combustion engine.

But the booklet[2] that came with the model dispelled that view. The engine is very much alive and has been the subject of serious developmental efforts in recent years by such prestigious companies as N.V. Philips of Holland, General Motors, the Ford Motor Company, and Stirling Thermal Motors, Inc. of Ann Arbor, Michigan. I'm indebted to Roelf J. Meijer and Ted M. Godett of the latter company[3] for valuable information.

Fig. 1. The model Stirling engine. For scale, the flywheels are 5.6 cm in diameter. Note the (white) wick of the alcohol burner.

In the final half of the 1800s and up until World War I, Stirling cycle engines were made and sold in large variety for powering things all the way from water pumps and boats to church organs and sewing machines. Today, it is doubtful that a Stirling engine, other than a toy model, can be bought commercially. But there are plenty of working examples in the laboratories mentioned. Ford has even built an experimental automobile powered by a Stirling.

A description of the model I purchased will give all the essential principles, but you should be aware that a design for a high-power engine would look quite different in configuration. The essential parts are shown in Fig. 2 at 90° intervals in a cycle. The engine is a closed system, containing a fixed amount of air. The air is free to move between parts of the system, so the pressure throughout is the same at all times, but the pressure does change from one part of the cycle to another. That's the result of some of the air being moved from where it is in contact with a cool surface to where it is in contact with a hot surface, and back again. The hot and cool surfaces are the two halves of a cylinder, as indicated in Fig. 2, the one half being heated by a flame and the other cooled by fins. The transfer of air from one of those parts to the other is accomplished by a large piston (closed at both ends) called a displacer, shown as D.

The displacer is smaller in diameter than the cylinder; it is kept on axis by the sleeve, S. As the displacer moves, air flows around it, so there is no difference of pressure on its two ends. It therefore does almost no work against the air; it only moves it. The alternating pressure, due to the alternate heating and cooling, does work against the power piston, P, and that is what makes the engine run. Let's see what happens at the four stages of the cycle.

First, note that the displacer and power piston move in quadrature; their crank pins on the flywheels are 90° apart in rotation. In the first diagram (0°), air is being heated. It is expanding and flowing around the displacer (indicated by an arrow), raising the pressure behind the power piston (as well as everywhere else). It is pushing the piston, which at that part of the cycle is in midstroke and moving with its maximum velocity.

At 90° (diagram at top right), the power piston is nearly at rest (just reversing), while the displacer is at maximum velocity, moving air from the hot part to the cool part of the cylinder, and cooling it in the process.

At 180°, the displacer is nearly at rest. The cooler air has lower pressure, allowing the power piston to return. The power piston is shown at midstroke, maximum velocity.

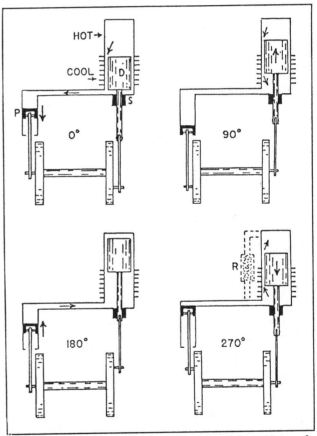

Fig. 2. Diagrams of the essential parts of the engine, at intervals of 90° rotation of the flywheels. The movements of the air and the pistons at each stage are shown by arrows and explained in the text.

At 270°, air is being transferred back to the hot part, while the power piston spends time at the end of its stroke. That gets it ready to start the sequence over again.

The final diagram shows how a very important improvement in efficiency is made, not one that is done in the toy model, but one that is standard in full-scale power engines. It is called a regenerator. The space for air to move around the displacer is eliminated, and instead it is sent around a bypass, as shown by dashed lines as R. The "can" in the middle of the bypass contains something like metal wool. The metal wool absorbs and retains some of the heat as the air moves out of the hot end of the cylinder and gives it back as the air moves the other way: a simple way to conserve heat, and effective.

It is said that if the Stirling engine is run by an external power source, it can act as a heat pump. It can bring heat into your house from the outside in winter or take heat out in the summer—or just cool your refrigerator. Out of curiosity I ran my model with an electric motor, first in the rotation direction in which it urns normally when heat is applied to the back end of the cylinder. Heat was removed from that end—it got cool, and the part with fins got warm. It the reverse rotation, the reverse happened. The temperature difference was small. A regenerator would have helped.

The Stirling engine has more characteristics in common with the steam engine than with the internal-combustion engine. Like the steam engine, it can run on anything that can make heat—even garbage. It is quiet. (Think how different city streets would sound if the motors were Stirlings!) It requires a heating-up time before it can start. Pollution can be low because fuel can be burned in the optimum way. As in the steam engine, the limit on efficiency is dictated by thermodynamics, specifically the Carnot cycle, which means that the hot end of the cylinder should be very hot. And to get a lot of power out of a small package, the pressure of the gas should be high. The Stirling can do things the steam engine can't do, such as act as a heat pump for heating or refrigeration, and it can run on solar heat; only a parabolic mirror need be added.

A little early history may be of interest. The engine was invented in 1816 by Robert Stirling, a Scottish minister who was also a classical scholar and scientist of some esteem. Mechanical talent seems to have been the rule in his family: his brother, as well as four of his sons, were engineers. Some 60 years earlier (1756), his grandfather invented the rotary threshing machine. In his patent, Stirling placed much emphasis on what he called an "economiser," later to acquire the name "regenerator."

When pollution (both noise and air) grows worse and burnable refuse becomes more plentiful and more of a problem, will we be driving around in Stirling cars? Perhaps.

References
1. Solar Engines, 2937 W Indian School Road, Phoenix, AZ 85017.
2. Andy Ross, *Stirling Cycle Engines* (Solar Engines, 1977). Contains a list of other publications and suppliers of models. For articles accessible in a library, see: *Sci. Am.*, Aug. 1973, and *Pop. Sci.*, Feb. 1973.
3. 2841 Boardwalk, Ann Arbor, MI 48104.

How Does the Honeybee Sense Polarization?

A couple of years ago Nathan Nichols of Western Michigan University[1] posed to me the question of how, in terms of physics, the honeybee and many other insects are able to sense the polarization of the light of the sky and use the result for navigation. He gave me what he had found, which left some important questions unanswered. As it quickly turned out, the problem lies not in finding writings on the subject,[2] but in finding out the physics. For guidance on that aspect I am grateful to Thomas Cronin,[3] who wrote to me in some detail and suggested several papers that might throw some light on the physics. The most current one in his list, as well as two that should be available in any library are cited.[4-6]

On receiving Nathan's letter, what first came to mind as a logical explanation was that the bee, in effect, wears a pair of Polaroid glasses. And as a qualitative idea, that was not far wrong, for recent research has shown that the discrimination in the bee's eye is accomplished, as it is in Polaroid, by long polar molecules aligned parallel to one another.[7] In both cases the molecules absorb the energy of that component of the light that has its (electric) polarization in the direction in which they are aligned.

So far the same; the difference lies in what happens to the light energy that is absorbed. In Polaroid it is wasted—turned into heat. The light that is *not* absorbed passes through, and that is what we see when we wear Polaroid glasses. Polaroid is a simple subtractive filter.

In the bee's eye the opposite is true. The energy absorbed by the aligned molecules is what does the business. It activates a series of chemical changes that end up sending a message along the optic nerve to the brain. The component not parallel to the alignment of the molecules is wasted – probably as heat.

The polarization-discriminating molecule has been termed a gate-keeper.[6] That's because its structure is reversible. When it absorbs, its structure changes so as to open a (molecular) gate, which results in a message to the brain. When the light ceases, it returns to its original structure – the gate closes.

The polar molecules in question are none other than what used to be called *visual purple*, but now called *rhodopsin*. We, and all creatures that have eyes, have it. It lies in a layer in front of the sensitive ends of the nerves that lead to the brain. So if we all have it, why don't we discriminate polarization? A simple reason. While in our eyes those molecules are individually polarization-sensitive, they are oriented in a mixture of directions, so there is no combined effect.

Polarization discrimination has been found and studied in many insects besides bees. And fortunately so, for doing experiments. Following a flying insect like the bee with instruments would have its problems. A land-going insect is more suitable. A species of desert ant has been a favorite subject. Much of what we know about the bee has been found out from that insect, not from the bee at all!

Rudiger Wehner,[5] who has done much of the field work, recounts his experiences in following the desert ant with a cart equipped with all manner of color filters, polarizers, and screens to cast shadows – for confusing the ant in calculated ways. He tells us that the polarization sensitivity is confined to the near-ultraviolet of the spectrum. Also, that of the 5,000 or so compartments of the insect's compound eye, only those in a limited area at the upper edge of each eye sense polarization – a fact shown by covering parts of the eye. In the rest of the eye the rhodopsin does not have the special alignment. In the polarization-sensitive areas, different parts are sensitive to polarization in different planes, so, in principle, the plane of polarization can be determined uniquely.

So much for the sensing ability. Let's see what the polarization pattern is that the bee looks at. Sky light is polarized by Rayleigh scattering:[8] scattering by particles smaller than the wavelength of the light, namely molecules of the air. The polarization is normal to the plane defined by the sun, the observer, and the point in the sky being looked at. The polarization is retained even when the light comes through clouds. The singular point in the sky for polarization is the location of the sun, as shown in Fig. 1. The diagram indicates also that the degree of polarization decreases toward the sun.

The polarization pattern contains the same information as would be gained by direct observation of the sun. Which leads us to ask why, when the sun is shining, the bee needs to use polarization. Evidently she[9] does. It has been found, by testing her under an artificial sun with no sky polarization, that she is not able to orient herself. The answer may be that doing it in alternate ways would require two "computer programs," so, since the polarization is the one that is always there, it is the one to use all the time.

How the bee processes the information from the polarization sensors to find her direction is not clear – at least not to me. One way (by analogy to modern pattern-recognition computer systems) would be to "input" the polarization pattern of the sky and then process that to find the singular point (the sun) or, say, the vertical line of symmetry. A way that would take much less "computer power," would be simply to scan the sky by waggling the body to find a null signal – in much the way we tune a radio.

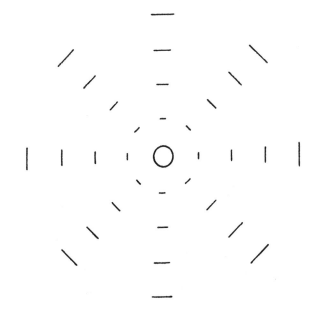

Fig. 1. The direction of polarization of the light around the direction to the sun, due to Rayleigh scattering. The lengths of the lines indicate that the degree of polarization decreases toward the location of the sun.

In whatever way the bee finds where the sun is, she still has to correct for the time of day. But that should be no problem. The insects all have built-in circadian rhythm[10] that regulates their activities.

Next time it's cloudy put on your Polaroids and see if you can find where the sun is.

References
1. Nathan Nichols, Physics Dept., Western Michigan University, Kalamazoo, MI 49008.
2. The first, and still the classical, treatise was written by Karl von Frisch over 40 years ago: *Bees: Their Vision, Chemical Senses, and Language* (Cornell University Press, 1950). This

and other earlier writings discuss mainly behavior, not the physics.

3. Thomas W. Cronin, Dept. of Biological Science, University of Maryland, Baltimore County Campus, Baltimore, MD 21228.

4. Rudiger Wehner, "The neurobiology of polarized vision," *Trends Neurosci.* **12**, 353 (1989).

5. Rudiger Wehner, "Polarized-light navigation by insects," *Sci. Am.* **235**, 106 (1976).

6. Conrad J. Mueller and Mae Rudolph, *Light and Vision* (Time-Life Books, Time Inc., New York, 1976), Chap. 4, "From Light to Sight."

7. See "How Things Work," *Phys. Teach.* **23**, 304 (1985).

8. J.W. Strutt (Lord Rayleigh), *Phil. Mag.* **41**, 107 (1871), or C. Bohren and A. Fraser, *Phys. Teach.* **23**, 267 (1985).

9. Feminine gender is standard terminology for the worker bees.

10. See "How Things Work," *Phys. Teach.* **27**, 470 (1989).

Newton's Yo-Yo, a Square Sprinkler, and a Disposable Battery Tester

As writer of this "column" I receive queries, hardware, suggestions for topics, and bits of information from readers and friends. I'm especially pleased when I receive a gadget to work with and try to figure out. That's true of the first and third items in the title above. You are invited to send things. I'll always try, with no promise that I can solve your puzzle!

Elizabeth Wood, author of the delightful paperback *Science for the Airplane Passenger*[1] sent me "Newton's Yo-Yo,"[2] a toy that would add pep to any lecture on elastic collisions. Figure 1 is a pen copy of the illustration on the package. Two plastic balls can be whirled around a spindle by rhythmic motion of the hand. They are tethered, each by two thin plastic rods, in such a way that they move in the same circle. That insures that collisions are centered— head on. And being elastic and of equal mass, the balls simply exchange velocity. One is red and the other yellow.

The instructions suggest two modes of motion. 1) The yellow ball is at rest while the red one goes the full circle, to hit the yellow one from behind. Red stops dead and yellow makes the circle, and so on. If whirled in the vertical plane, the collision point remains fixed near the lowest point of the circle. When the whirling is fast, all one sees is an apparent rapid alternation of the color of the station-

ary ball. 2) In the other mode the balls move symmetrically, each in a half circle, colliding at two points diametrically opposite. It takes more practice to start it going that way, and it seems to be stable only when the frequency is high—5 to 10 collisions per second.

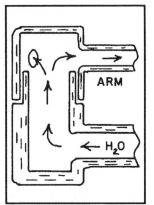

Fig. 2. Simple version of a rotating sprinkler, in cross section, showing how water flows from the stationary base into the arms.

Brad Donahue, a physics student at Glenbrook North High School,[3] is puzzled about a rotating sprinkler that sprinkles a square patch of lawn. To begin with, a little examination of how an ordinary rotating sprinkler works may point to a simple modification that can make it sprinkle in a square. The key lies in how the water gets from the stationary base to the rotating top. The construction varies, but in simple-minded concept it is as shown in Fig. 2. The sprinkling arms (usually three) extend out from an inverted cup that sits over a short pipe coming up from the base. The cup and arms rotate. Now imagine that the open end of the pipe from the base is not the same height all around, but is "scalloped," that is, higher in four places, at 90° intervals. High enough in those places to cover partly the hole through which water exits to the passing arm. So in those four directions the squirt will be less, making the flat sides of the sprinkled square. I examined a square-pattern sprinkler[4] at a garden store, but did not dare to take it apart. Its outside appearance confirmed the above explanation.

Leonard Jossem,[5] a former AAPT president, sent me an unusual battery tester with a challenge to figure it out. It is built into the display package that contains four AA-size alkaline batteries.[6]

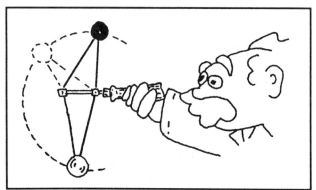

Fig. 1. A pen copy of the illustration on the package of "Newton's Yo-Yo."

The tester, one sees right away, is made with the familiar "liquid crystal," a material that changes its appearance at some critical temperature. It does that because its molecular order, and consequently its transmission and scattering of light, is a rather sharp function of the temperature. In a common form the change takes place between room temperature and body temperature so that holding a finger on an area of it leaves a temporary image of the finger.[7]

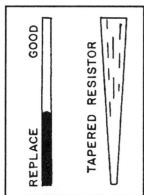

Fig. 3. The parts of the battery tester. On the left is the liquid crystal column with labels, which is on the front side. The tapered resistive coating, which is behind the liquid crystal column and in thermal contact, is shown on the right.

The tester Leonard sent looks more like a thermometer than a meter. A narrow vertical column of the liquid crystal (in thin plastic of course) is in front of, and in thermal contact with, an electrically conducting film. These are shown separately in Fig. 3. The film is tapered, narrow at the bottom of the liquid crystal column and wide at the top. So the resistance per unit length is tapered likewise — highest at the bottom.

In total, top to bottom, it is about 10 Ω.

The battery to be tested is easily inserted from the rear of the package to make contact with clips at the ends of the tapered resistor. The temperature rises in the resistor according to the resistance-taper and comes to equilibrium in a few seconds. If the battery is good, the whole tapered resistor, up to the widest end, will reach the critical temperature, so the liquid crystal will change appearance up to the top of the column — where a label reads "good." If the battery is nearly spent, only the narrow part of the resistor will get warm enough, and the label there says "replace." There seems to be no correction for ambient temperature — evidently indoor use is expected.

References
1. Ballantine Books, New York, 1968.
2. Fascinations, Inc., 309 S Cloverdale, Suite B3, Seattle, WA 98108.
3. 2300 Shermer Rd., Northbrook, IL 60062.
4. The Nelson Rainswirl #45, "Poppy" square-pattern sprinkler. Available in garden stores, e.g., Frank's.
5. Physics Department, Ohio State University, Columbus, OH 43210.
6. Duracell batteries: the package says "New! Coppertop tester."
7. Liquid crystal sheets, having various critical temperatures, are listed in the Edmund Scientific Co. Catalog.

What Makes a Fishing Spinner Spin?

More than two years ago Jon Roberts (student) and Dean Hecht (teacher) of Superior Senior High School[1] asked why a fishing spinner spins.

For readers who fish, a description of a spinner is not needed. But for others a few words may be welcome. The part that spins is a metal spoon, just about the shape of a teaspoon or tablespoon, but not as deep, and it may be from 1 to 5 cm long, or even more, depending on the size fish to be attracted. The spoon is linked at its upper end to a short stiff wire, in such a way that it is free to move in any direction. Hooks dangle at the end of the wire. In some cases colored beads are strung on the wire above the hooks to better attract the fish in the water—or the fisherman in the store. A typical spinner is sketched in Fig. 1A.

When being pulled through water, the spoon goes around and around the wire at more than 10 rps, while maintaining an angle of 30° or 40° to the wire (Fig. 1B). For reasons known only to a fish, it appears as something to be attacked or eaten.

The question asked by Roberts and Hecht is why a perfectly symmetrical spoon should be propelled around the wire at a high rate by the water. To that I will add another: Why should it move out to such a large angle from the wire while being pulled along?

The writers had done some investigation on their own. Their co-workers, students Joe Depta and Brad Till, had taken flash pictures of spinners being pulled through water, one of which is shown as the left side of Fig. 2. They sent several spinners for my own observation. Not that I needed the introduction, having fished with them many times.

I pulled the students' spinners through water, but could see no more

than their pictures showed. Not having an underwater camera, there seemed to be no way to get any other views. So the matter was put on the shelf. Recently the thought occurred to me that something might be learned from a simulation in an air stream, where flash pictures could be taken of all aspects. That was tried; it gave some enlightenment and some surprises, but certainly not the full explanation for a fishing spinner in water. By now the students may have graduated and left, but here's what I found anyway.

Fig. 1. A fishing spinner. A shows the shape of the spoon. B shows the outward slant of the spoon when it is spinning and the way it is linked to the central wire. The line is tied to the ring at the top.

To make a parallel stream of air, I mounted a 12-bottle cardboard carton (with bottom removed, but bottle separators retained) on the front of a 12-in electric fan. The spoon, 15 cm long and 9 cm wide, was cut from moderately heavy paper. To make it spoon-shaped, it was slit on its line of symmetry and taped back together. It was tethered to a thin brass rod, concave side toward the rod, by a small wire loop at its top end. In the air stream from the fan, the spoon

went around—in whichever sense it was given a start—fast enough to be a blur. Flash pictures were taken in several aspects, one of which is shown as the right-hand part of Fig. 2. The photograph also shows the arrangement of the fan, the multi-channel carton, and the rod on which the spoon is tethered. The figure caption identifies the details.

Mistaken idea #1. I had assumed that in order for the spoon to be pushed around the wire it would have to maintain a position such that the rod was not in its plane of symmetry. To try to see that, many flash pictures were taken from the side with the assembly turned so that the rod pointed vertically downward in order to eliminate possible asymmetry due to gravity. The rod appeared to be in the plane of symmetry of the spoon.

The result was a little hard to believe. So a check was made in a sort of reverse way; that is, to *force* the spoon to remain in the symmetrical aspect and see if it would still spin. At the center of the concave side of the spoon a wire was attached so that it stuck out in the plane of symmetry and looped around the brass rod. It was braced to the edge of the spoon, as seen in Fig. 3A. Its length was such that it fixed the outward slant of the spoon at about 30°. The spinning was as fast as before. It went equally fast when started in either sense, which showed that the constraining wire was in the plane of symmetry of the spoon as intended.

Mistaken idea #2: I had assumed that the convex surface of the spoon is a necessary feature, having some airfoil effect like the lift of an airplane wing. So a perfectly flat blade was tried. It spun as well as the dished one when started either way and with or without the constraint described in the paragraph above. Then why do the fishing spinners have spoons, not flat blades? Better self-starting, mechanical rigidity, focusing of light?

The $64 question. Needed is a moment of force about the axis of rotation to act on the symmetrically oriented spoon. I'll give my best guess, but keep in mind it is about a spinner in air with specified orientation, not the real one in water. We might look to see how the pressure of the air varies over the surface. When the spoon is spinning and being pulled forward, the air approaching its outer side has a component of velocity toward the surface because of the spoon's outward slant, and a transverse component because of the spoon's rotation around the rod. In the plane to which the rod is normal, the combined velocity is indicated in Fig. 3B.

If you remember my discussion about sailing tin can lids and Frisbees,[2] you know that when air is deflected by a surface, the center of pressure almost never is at the center of area. So we guess that in Fig. 3B the net force of the air on the spoon (including both inner and outer surfaces) is as shown by the downward arrow, which passes the center of rotation (rod) to the right. That gives a moment of force in the right sense to drive the rotation. Now before you say that is working backwards from the answer, I'll admit it. But it's the best that can be done. Centrifugal force, which probably accounts, mainly, for the outward slant of the spoon, gives no moment to confuse the issue because it acts from the center of rotation.

At the end I can only say, "Whew! Those fellows at Superior Senior High School didn't know what a hard problem they were posing!"

References

1. 2000 Catlin Ave., Superior, WI 54880.
2. "How Things Work," *Phys. Teach.* **21** 325 (May 1983).

Fig. 2. Left side: flash photograph of a spinner being pulled through water, taken by the group at Superior Senior High School. Right side: spoon made of light cardboard, spinning in an air stream. For visibility in the flash photograph, the edges are white, a white stripe identifies the plane of symmetry, and a piece of white tape identifies the outer end of the rod.

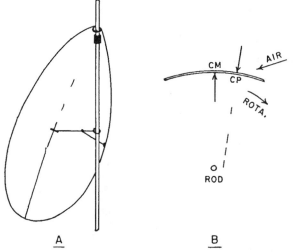

Fig. 3A. Sectional view of the spinner in a plane to which the rod is normal. 3B. Center of mass, center of pressure, direction of attack of the air, and the sense of rotation are indicated. The downward arrow indicates the effective force of the air, and the upward arrow the centrifugal force.

Digital Electronic Balances: Mass or Weight?

My first exercise in chemistry lab, long ago, was to weigh something on an analytical balance. To achieve the final precision, small squares of aluminum foil (in the milligram range) were put on and off with tweezers, and if no combination gave an exact balance, an interpolation could be made by reading the pointer or by sliding a small mass along the beam. Quite a lengthy process, but good discipline.

Nowadays in many labs, weighing scales that have a single pan and a digital readout replace the traditional ones that have a beam, two pans, and a pointer. The new kind works in quite a different way. It has a beam, with a pan at one end that holds the object to be weighed. At the other end of the beam a balancing downward force is supplied by a current-carrying coil of wire in a constant magnetic field. The weight of the object on the pan is determined from the current in the coil that just produces balance. That's the general idea.

I visited the local shop[1] that repairs digital balances and saw the insides of several different ones. Dave Bliss and Scott Kureth explained the details of their operation. Also I received valuable information from William J. Halsey of Mettler Instrument Corporation.[2] The rather complicated construction of such balances can be reduced to a simple schematic diagram, as in Fig. 1.

The coil (C) is in the field of a permanent magnet (PM) in much the same arrangement as is seen in a cone type loudspeaker. The coil is a hollow cylinder that can move up or down in the cylindrical gap of the magnet. It pulls downward with just enough force to keep the beam horizontal for any load that may be in the pan—within limits, of course. The current required to do that is controlled automatically. The first question to ask is how the current is controlled.

The circuit is a feedback loop. When the beam is horizontal, a "knife-edge" (K) cuts off part of the light that is passing from a source on one side to a receiving "eye" on the other side (S and E). The eye is a photosensitive diode, whose resistance is a function of the light incident on it. Its resistance controls the current that an amplifier (A) sends through the coil. The electrical feedback path is indicated by the wavy lines with arrow heads. If the beam starts to tilt, say clockwise, letting a little more light through from source to diode, the current in the coil increases just enough to keep it from tilting further. So equilibrium is maintained where the beam is very nearly horizontal.

The current that keeps the beam horizontal is read out on what basically is a digital ammeter (DA). But of course the display reads not in amperes, but in milligrams or grams.

A question arises that did not have to be asked about the conventional weighing balance. What if the digital balance is taken to a place where g is different—the Moon? Or just moved from pole to equator on the Earth? We don't ask that about the conventional balance because, although it really compares the weights (forces of gravity, mg) on the two sides of the beam, the same comparison goes for the masses. That's because g is the same on the two sides and cancels. In fact, the pieces of brass and aluminum that are put into the pan as standards are stamped with numbers that state masses. Not so simple with the digital device. It compares mg on one side with an electromagnetic force on the other, and the latter does not change along with g. So, as described up to this point it measures weight, its closest relative being the old-time spring weighing scale.

Of course there is a way around the problem. Digital balances are provided with a standard mass that can be lowered onto the pan by the push of a button or turn of a knob. Then an adjustment is made in the digital-reading ammeter part of the circuit so that the display shows the value of the test-mass that is on the pan. The adjustment is such that linearity is preserved, and the display still reads zero when the test-mass is

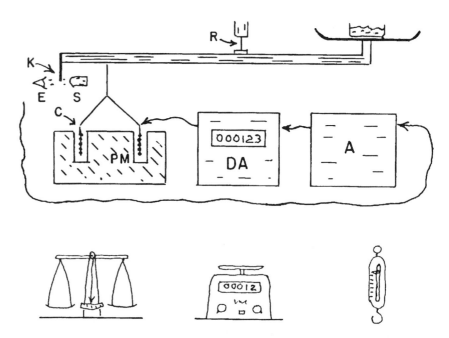

Fig. 1. Above: A much-simplified version of an electronic digital balance. A coil in the field of a permanent magnet pulls down on one end of the beam to oppose the weight of the object in the pan at the other end. The current in the coil is controlled by a feedback loop to keep the beam horizontal. The mass of the object is determined from that current. Below: Reminders of the three classes of weighing devices discussed in the text: two-pan, electronic digital, and spring.

lifted. Not a hard electronic problem, but not one to be gone into here. Thus after calibration with the test-mass, the device, in effect, reads mass and would do so on the Moon if calibrated there. Expensive digital balances have automatic calibration at the press of a button.

There is another difference between the modern digital balances and the conventional ones, as I learned from Scott Kureth. It concerns the support of the beam. The beam of the conventional balance is supported, typically, on a knife edge, and the pans are hung from the beam by knife edges. Even though the parts are made of the hardest steel, or sapphire, there is finite friction, and that places a lower limit on the amount of unbalance that will cause the beam to move. The beam in a modern digital balance is supported instead by a thin metal ribbon or the equivalent (R) that flexes as the beam tilts. With such a support there is no lower limit to the unbalance that will cause the beam to move. If such a support is flexed a little when the beam is in the neutral position, giving a small torque, that gets taken into account in the calibration.

A point about accuracy and errors. The value of g varies due to the rotation of the Earth, for example from Reykjavik at 64° N to Quito at the equator by half a percent. The effect of a difference in altitude at a given latitude is far less. Changes due to g are well above the sensitivity limit of the balance, which, according to one figure, is 0.01 mg in a range of 0 to 100 g. So calibration at the place of use is essential. Temperature variation, causing dimensional changes in the parts of the balance, is the greater enemy of precision weighing. Even though internal temperature compensations are built in, calibration at the time of use is necessary if high precision is required.

Digital balances are made having ranges of zero to a few grams, up to zero to tens of kilograms. The widely different ranges are obtained, mainly, by choice of the ratio of the lengths of the arms of the beam.

References
1. Toby's Instrument Shop, 2370 Abbot St., Ann Arbor, MI 48103.
2. Box 71, Princeton-Hightstown Rd., Hightstown, NJ 08520.

The Rattleback Revisited

With all that has been written, during nearly a century, about the rattleback, one might think there is little left to be said. But it hasn't ceased to puzzle and intrigue. I'm one of the many who have looked for natural rattlebacks in streambeds or have made experimental ones. I think I can add a little to the lore.

Rattleback is a name coined by the late A.D. Moore[1] for the object that earlier was called a celt. It is a solid object (most often a stone) of a particular shape that will, if spun on a flat surface, come to a stop, "rattle" for a moment, and then proceed to spin in the opposite sense. It does that only if started in one sense; in the other sense it just spins until it comes to rest. Some pocketknives will exhibit the behavior if spun closed with the blade-side up. Selected smooth stones, picked up from a streambed, will show the behavior. Nowadays plastic versions can be purchased in toy stores or from catalogs, under various names, for example "Space Pet."

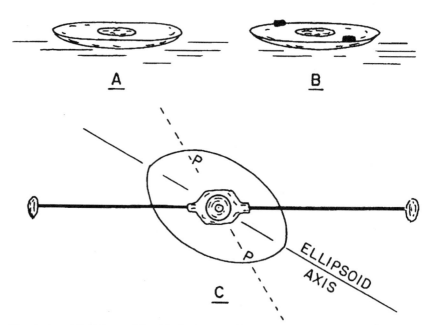

Fig. 1. A and B: Slices of hard-boiled egg with shell, without and with asymmetrical masses added. C: View looking down on a rattleback, showing the sensitive points P, located symmetrically with respect to the arms about the ellipsoid axis.

Celt is the name of a prehistoric stone artifact[2] probably used as an axe head. Evidently because some of them reverse on being spun, the name has been adopted for any object that reverses.

In the course of experimenting to find the best shape, A.D. Moore made hundreds of rattlebacks[3] by casting them in dental cement, the kind the dentist fills your mouth with to take an impression of your bite. A.D. (as he liked to be called) told me that his best ones would reverse more than once after a spin. The examples I have tried have done that not at all or only feebly.

The requirement for an object to show reversal of spin is simple enough. It has been explained up and down as far back as 1896.[4-9] The surface in contact with the table must not be spherically symmetrical. Commonly it is approximately ellipsoidal. The distribution of mass must be unsymmetrical with respect to the axes of symmetry of the ellipsoid. In just *how* the mass is distributed, unsymmetrically, lies the art of making a rattleback that will perform well. Undoubtedly that was the main variable with which A.D. experimented.

To illustrate the requirements, consider a slice of hard-boiled egg with shell intact (Fig. 1A). It will not reverse. But if mass is added unsymmetrically as in Fig. 1B, it will do so. Often these requirements are satisfied by smooth stones found in a streambed: ellipsoidal surface and unsymmetrical mass distribution.

The so-called rattle, which occurs just as the rotation comes to a stop and goes into reverse, is mysterious only because the motion in that brief time is too rapid to follow. There being no rotation, the motion must be of a rocking mode. And it must be storing the energy because it will again rotate. To see just what is happening, we need to slow the motion down. That is easily done by locating the unsymmetrical, added mass much further from the point of contact with the table. I have experimented with that and have ended up with added

Fig. 2. An assembled rattleback, the forms used for casting the plaster of Paris, and the crucible for making the solder blobs.

masses at the ends of stiff wires or rods as can be seen in Fig. 2. The height of the masses (adjusted by bending the rods) is such that the device will rock a few times per second. The plane of the rods with respect to the plane of symmetry of the ellipsoid is set (by the nut at the center) to give the best reversing.

Now the motion can be followed easily. If the gadget is started at a few revolutions per second, it will come to a stop, rock at a few cycles per second, and resume spinning in the opposite sense.

Something interesting can be seen. The points P of maximum up-down motion appear to be on the (dashed) line, symmetrical with that of the rods (Fig. 1C). That suggests a different way of starting the motion. With the device at rest, if at either point P in the figure a sharp downward poke is given with the finger, rocking motion starts, and then converts to rotation. If the poke is given at a place other than P, the rotation is less.

Brian Pippard[10] and I have discussed by correspondence a modification that might be entertaining. That would be to make the device big enough so a child can sit in it. By rocking it (no problem

for the child), rotation might be started and driven. For anyone who wants to try that, there is an easy way of getting a boat-like ellipsoid. The plastic or aluminum dishes that kids use for sliding on the snow are the right size. Deforming one from spherical shape to ellipsoidal might not be hard. For unsymmetrical loading, masses could be clamped to the rim, or put on "outriggers." Any takers?

Here are brief notes on making the rattleback of Fig. 2. Over the years I have made many. Once the simple implements are gathered, production is fast. Make one for a friend who might need an "executive tranquilizer!"

The picture shows an old tablespoon. It is pressed into a piece of modelling clay (shown below it) to make a form in which to cast plaster of Paris. After the plaster has hardened, the clay is easily bent a little and pulled free.

A screw, threaded end pointing upward, is cast into the plaster. To make sure it doesn't work loose, a small crosspiece is soldered to its head. The plaster crucible shown is used for forming solder blobs on the ends of the arms. Solder is melted in the crucible, and the end of the brass rod or wire, with flux, is held in it while the solder solidifies. The plate under the nut, to which both arms are soldered, is made of sheet copper that can be easily bent to change the upward slope of the arms. That determines the rocking frequency. The nut allows adjustment of the azimuth of the arm assembly, which is set for maximum reversing action.

References

1. A.D. Moore, Department of Electrical and Computer Engineering, University of Michigan, unpublished paper, 1972.
2. *Encyclopaedia Britannica* article on the celt.
3. J. Walker, "The amateur scientist," *Sci. Am.* **241**, 172 (1979).
4. G.T. Walker, *Q.J. Pure Appl. Math.* V, 28 (1896). The first analysis of the reversing celt.
5. H. Crabtree, "An elementary treatment of the theory of spinning tops and gyroscopic motion." First published in London in 1909. Reprinted in 1923 by Longmans, Green, New York, and in 1967 by Chelsea, New York.
6. A. Sommerfeld, *Mechanics* (Academic Press, New York, 1952), p. 149.
7. A.J. Boardman, "How to make celts of wood," *Fine Woodworking*, July/Aug., 68 (1985).
8. H. Bondi, "The rigid body dynamics of unidirectional spin," *Proc. R. Soc. London, Ser. A* **405**, 265 (1986). An exhaustive treatment.
9. A.B. Pippard, "How to make a celt or rattleback," *Eur. J. Phys.* **11**, 63 (1990).
10. Cavendish Laboratory, University of Cambridge, Cambridge CB3 0HE UK.
11. Plaster of Paris from any art supplies store; clay from same or a toy store.

Steam Clock in Gastown

Most attendees at the 1991 summer AAPT meeting in Vancouver, British Columbia, will agree that watching the steam clock and hearing its steam whistles toot and play the tune of the Westminster chimes was one of the week's more memorable experiences. I spent the better part of an hour there on a visit years ago.

Anticipating the interest of the visitors during this year's meeting, I corresponded with the inventor-builder, Raymond L. Saunders, a noted horologist and operator of a clock business in Vancouver.[1] When I asked him for some plans and technical specifications, he said there were none. That's typical for something that is done for the first time. At least it is so for me, when I start something new. But now that the clock has become famous, there is no lack of articles written about it by others. Ray sent me copies of several, two of which are readily accessible.[2,3]

A fact to start with is that live steam is piped under the ground in parts of Vancouver for heating purposes. The story goes that in the early '70s the city planner of Vancouver told Ray, whom he knew to be a clock-maker, that he was looking for a way to cover a steam vent at the corner of Cambie and Water Streets with something that would make use of the steam. That was soon after the area, formerly a "skid row" known as Gastown, had been rehabilitated into an attractive restaurant and shopping center. Ray took the challenge, and the result was his steam clock, unveiled in 1977.

Figure 1 is a picture of the steam clock made from a color positive loaned by Ray who stands beside his creation, dressed especially for the photograph. Note what appears to be a cloud of condensed steam, at the upper left.

The clock is 16 ft (4.9 m) tall and weighs about 2 tons (1800 kg). On each of the four sides there is a clock face at the top and a window below for viewing

Fig. 1. The Gastown Steam Clock, with a cloud of condensed steam at its top, being shown by its inventor/builder Raymond Saunders, who dressed up especially for the photograph.

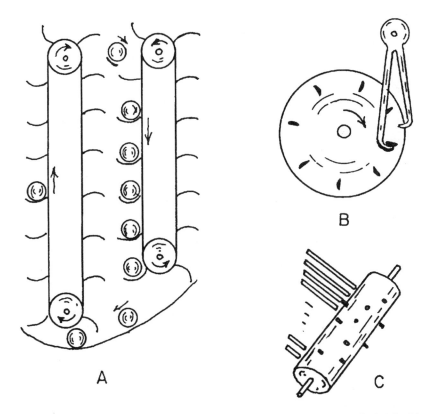

Fig. 2. A—The driving mechanism for the clock. The endless chain on the left is driven by the steam engine. It raises balls to the top and delivers them to the right-hand chain. The balls on the right-hand chain move downward under gravity, producing the power to drive the clock. B—A pinwheel escapement. The disc is driven clockwise. The pendulum stem is attached to the top of the inverted V-shaped piece and extends downward to the bob. C—Sketch showing the idea of the old-time music box. In modified form, a cylinder and "finger" arrangement controls the steam whistles and causes them to play their tune in the Vancouver clock.

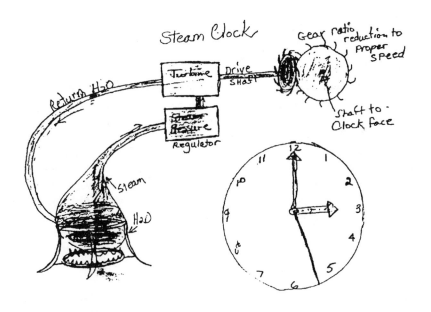

Fig. 3. The concept of a steam clock submitted by a young person in the Science-By-Mail program.[5]

the mechanism. The clock is of the pendulum type, powered, as is the conventional grandfather clock, by a falling "weight." But this falling weight is one of a kind, as we shall see. The swinging bob, gold-plated and of 20 kg mass, can be seen through the windows.

For the physicist, there are several interesting parts to the Gastown clock. The first is the steam engine. It has a single cylinder and runs continuously. It has to run continuously because of a characteristic of all one-cylinder steam engines: in each cycle there are two positions at which there is no torque. If it stops at one of those positions, it will not restart by itself. (If you are old enough, remember that a steam railroad locomotive has two cylinders. The two work with a phase difference of 90 degrees, so when one does not produce torque, the other does.)

The main job of the steam engine is to raise weights, which then fall under gravity to supply the power for the clock. But with a difference. In the standard grandfather clock the weight, of a kg or more, is raised through its full distance once a day—in some cases once a week. In Saunders's clock, the steam engine raises a little bit of the weight from bottom to top every 4.5 minutes. So it is fully "wound" all the time.

The continuous process is accomplished by two endless chains that carry, in wire loops or scoops, steel balls (see Fig. 2A). The one on the right has one side filled with balls, which move downward under gravity, producing power that is taken off to the clock from the top sprocket wheel. The chain on the left is driven, continuously, by the steam engine. Each time a ball on the right chain (the power chain) reaches the bottom, it rolls off and is picked up by a loop on the up-going side of the left-hand chain. It rides up and is transferred to the top of the power chain. A ball is delivered every 4½ minutes.

It's interesting to see what power is required of the steam engine. In J/s it is the weight of a ball (5 N) times the height raised (2 m), divided by the time (270 s). Neglecting friction loss in the chain, the steam engine has to deliver only 0.0037 J/s or 5×10^{-6} horsepower.

A toy steam engine would be powerful enough!

The clock escapement is one of the less usual ones,[4] a "pinwheel," as sketched in Fig. 2B. The flat disc carrying the pins is driven by the power from the ball system, through the gear train. The pendulum stem is attached to and extends downward from the top of the hairpin-shaped piece that has the pallets (hooks) at its ends.

The blowing of the steam whistles is controlled by a relative of the old-time music box. In that device there is a rotating cylinder from which pins project (Fig. 2C). The pins twang steel reeds that vibrate at the frequencies of the different musical notes. The arrangement of the pins on the drum determines the tune that is played. The steam whistles are controlled in the same way, except that instead of the reeds there are fingers that actuate microswitches. When one of them closes a circuit, a solenoid valve lets steam into a particular whistle. That is the only part of the system that works by electricity. The cylinder is of course turned by steam power.

At 90-degree intervals on the cylinder there is a set of pins that produces the Westminster chime tune on the quarter hour. Four whistles are required for that. Then there is one set of pins on the cylinder that blows a fifth whistle on the hour. That whistle is a "king-size" one, duplicated from one that was salvaged from an old ferry boat. When the whistles blow, especially at the hour, great white clouds of steam rise.

Saunders has two more clocks under construction. One, a lunar clock in Bridgepoint Market in Richmond, B.C., a suburb of Vancouver, is 65-ft (20-m) tall, has a 6-ft (1.8-m) diameter moon disc, and a pendulum 42-ft (13-m) long. Another, the Sinclair Tower clock in Vancouver, has an antique mechanism weighing two tons, which gongs a 90-cm-diameter, 1400-kg bronze bell. So far Ray has not added whirling or dancing marionettes, but I won't be surprised if that's in his future!

As a matter of interest, there appears to be only one competing design for a steam clock (Fig. 3). That's one dreamed up by Shoshanna Ebersole and submitted to the Museum of Science, Boston, in its Science-By-Mail program.[5] It is reproduced here by permission of its author and the Science Museum.

References
1. Gastown Steam Clock Co. Ltd., 123 Cambie St., Vancouver, BC V6B 4R3, Canada.
2. Ivan Slee, "The Gastown steam clock," *Model Engineer* **163**, 3859, Oct. 20, 1989, p. 506.
3. Mike Edwards, "Toots tell time to steam lovers," *Natl. Geogr.* **154**, 4, Oct. 1978, pp. 475–477.
4. The pinwheel escapement is not often seen in watches or clocks. The clock escapement has endless variations: in one reference book 18 are shown, most named after the inventor.
5. Shoshannah Ebersole, Entry #340, *The Science-By-Mail Post* **II** (1), 1990, Museum of Science, Science Park, Boston, MA 02114.

Three Intuition Teasers

Recently an advertisement[1] came to me showing a weighing balance that balances for all distances of the weights from the pivot. From a few sticks I put one together. See it in Fig. 1. The arms do not extend out from a simple pivot, but from the sides of a parallelogram. They stay horizontal as they move up and down, as seen by comparing the two views in the figure. There are a couple of ways of understanding why, if the weights are equal, they balance no matter where they are on the arms. One way is to note that if one weight goes down, the other goes up by the same distance, so there is no change in gravitational potential energy. The other way is to consider the lever arm for each weight to be the horizontal distance from the center to the side of the parallelogram.

The balance is more than an interesting puzzle: it demonstrates a principle of wide application. I compared it with the postage balance on my desk. Replace the arms (in Fig. 1) with pans supported on the two vertical sides of the parallelogram and you have it. Since those two sides remain vertical, the pans remain horizontal as they go up and down. Therefore it makes no difference where, on a pan, the load or standard weight is placed—a necessary condition for practical use. If you keep your eyes open, the parallelogram will be seen as the essential element in all balances whose pans are supported from below rather than by suspension from a beam above.

If you should set out to reproduce the balance for demonstration purposes, there are a couple of fine points to be observed. One of the horizontal members of the parallelogram should be constrained for horizontal motion only, which means that the pin should go through a vertical slot in the supporting post, as can be seen in the photographs, rather than through a hole. If both of those pins were to go through holes, they might pull against one another, and so increase the friction. Engineers would call that an "overdetermined" structure. The other point is that those same two pins should be located a little above the lines connecting the outer pins, so that without load, gravity will bring the parallelogram to rest with its long sides horizontal.

The second "teaser" is a class of rollers that invites one to experiment. Three

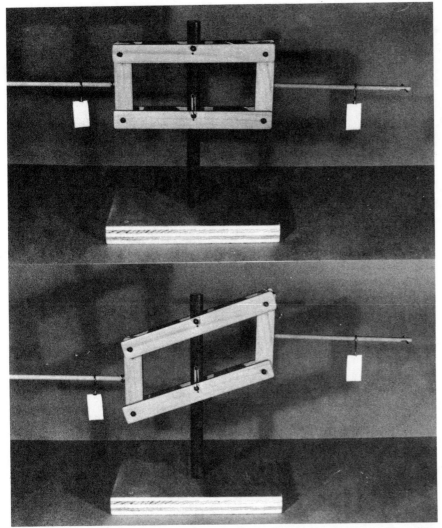

Fig. 1. The two-arm balance made with a parallelogram at its center. Comparison of the two views shows that the arms remain horizontal as they move up and down.

Fig. 2. Rollers having 3, 4, and 5 edges and sides. If the board is moved horizontally over the 3- or 5-sided roller, it stays at constant height above the table.

are shown in Fig. 2. If the board shown resting on the three-edged one is moved horizontally, it does not move bumpily; it moves smoothly at constant height above the table. The same is true for the five-edged one, but not for the four-edged one, shown standing on end at the far left.

The recipe for making a roller is simple: a side opposite an edge forms part of a circular cylinder whose axis is that edge. The edges are uniformly spaced, and all the radii (edge to opposite surface) are the same. (The requirement that there be a surface opposite an edge means that there must be an odd number of edges and surfaces—$2n + 1$ symmetry. The four-edged roller shown does not satisfy that.) To construct the three rollers shown, first make a pattern of cardboard, then cut a number of pieces from plywood. The pieces are stacked with glue, sanded, and varnished. They are of pleasing enough appearance to serve as ornaments on the desk or coffee table, with the added advantage that they may elicit a leading question.

One is tempted, first off, to make a cart having wheels of the shapes described. It would give a bumpy ride. There is no location for an axle such that it would stay at constant height above the table, as one will find with a little trying.

Interesting variations on the roller of three-fold symmetry are described in the *Exploratorium Science Snackbook*.[2] It does not go further in $2n + 1$ symmetry. It describes an easy way to make a roller. Two of the shapes are cut from cardboard and nailed or glued to the ends of a stick.

The third device to puzzle over is a wheel that appears as though it ought to go around and around all by itself. I saw one in the Deutsches Museum in Munich, Germany, many years ago. I've forgotten the title, but it might have been "Why isn't this a perpetual motion machine?" A reproduction of the wheel is shown in Fig. 3. It has hinged blocks around its periphery. The hinges are at the edges of the blocks, best seen in the one that hangs down at the bottom of the wheel. The picture on the left shows the wheel in one position, and the one on the right shows it after it has rotated through a small angle counterclockwise, as indicated by the white spot. The rotation has caused the block with the black spot to fall into a nearly horizontal position. By that process, the centers of gravity of the blocks on the left side of the wheel are at all times at greater radius from the axle than are those on the right side.

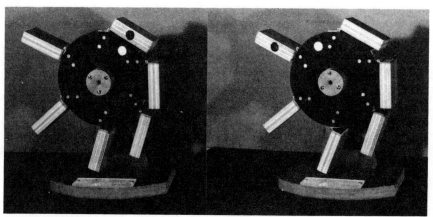

Fig. 3. A wheel with hinged blocks on its periphery, which appears as though it should rotate counterclockwise.

Their moments of force are greater. So the wheel should rotate, by itself, counterclockwise. Yes?

Of course physicists know perpetual motion will not occur, but it is fun to hear their explanations of why the sum of the moments of force of the six blocks will not produce a net torque on the wheel.

References

1. Educational Equipment Co., P.O. Box 2102, Vernon, CT 06066-2102. Item ME-150 "Balance Paradox."
2. *Exploratorium Science Snackbook*, Item 71. The Exploratorium, 3601 Lyon St., San Francisco, CA 94123.

Sensing the Rotation of the Earth

Have you ever stopped to think how many ways there are to sense that the Earth is rotating, without looking at the Sun, Moon, or stars? Or have you challenged a class to come up with such ideas? I got to thinking about that when looking back at an article in this journal by Mark M. Payne,[1] in which he describes seven of the ways. I've been a longtime collector of such ideas, probably as a result of my experiments with the Foucault pendulum[2] (which is one of the examples in Payne's list). I will add a few that I have collected. First, a brief rundown of Payne's list of seven.

First is the circulation of the wind in the two hemispheres. Second (which I find imaginative but a little far fetched) is the fact that we speak English. For the third, Payne cites the drift to the right of projectiles fired at London from Germany during World War II. Fourth, the turning of the Foucault pendulum. Fifth, the change in g at sea level from pole to equator. Sixth, the deviation to the east of an object dropped from a tower, and seventh the fact that the axis of rotation of a gyroscope does not remain fixed with respect to the Earth.

My first addition is one told to me a long time ago by my friend Robley Williams.[3] He related that once he happened to be staying in a hotel where a convention of railroad people was going on. Out of curiosity he slipped into one of their meetings and listened to a few talks. An observation they described as not being understood, he said, was that on the line between New York and Washington, which has a track for each direction, the east rail of the northbound track gets more wear than does the west rail, and vice versa for the southbound track.

Robley had not imagined that he would enter the discussion, but the mystery he heard described was too much. He proposed that the effect might be due to the rotation of the Earth: that in the northern hemisphere the moving train would lean to its right, putting extra load on the right rail and increasing its rate of wear. He advised the railroaders to look for the effect in the southern hemisphere, where the train should lean to its left. The railroaders listened politely, but with little reaction.

As you will recognize, the phenomenon is a variation on one sometimes demonstrated in a physics course.[4] A student sits on a rotating platform and launches a projectile aimed at a target that is on the same rotating platform. The projectile misses the target. In the frame of reference of the platform, it travels in a curve. The train, if it were moving freely on the rotating[5] Earth, likewise would follow a curve, as traced on the surface of the Earth. But it is restrained from curving by a track that is straight. To restrain it, the track has to exert a horizontal force on the wheels, as suggested by f in Fig. 1A. Since that acts below the center of mass, the result is that the train leans, as if to roll over, making the downward forces on the rails unequal. Robley had the physics right, but whether the effect would be large enough to account for what the railroaders thought they saw was not answered.

Victor Neher[6] has written to me about a method he understands was dreamed up at, of all places, the Vatican. It's a version of the demonstration in which a student sits spinning in a swivel chair, pulls masses inward that are held in the hands, with resulting increase in angular velocity. Two sticks or metal pieces are pivoted on a vertical spine, which is suspended by a fine wire (Fig. 1B). They are brought to rest in the local frame of reference, which means that they are then rotating with the vertical component of the Earth's angular velocity. At the latitude of Pasadena, where Victor experimented, that is 8.4 deg/hr.

Next the sticks are rotated to the vertical, as in Fig. 1C. The moment of inertia of the combination decreases by a factor that can be several hundred, thereby causing the angular velocity to increase by the same factor. For observing the rotation, a small mirror is attached at the pivot point, and a light beam is reflected from it to make a spot on a scale on the wall. Nowadays, the beam would be from a laser.

The practical points are interesting. The sticks were brass, of rectangular cross section, 25 cm long. The problem of reducing spurious rotation prior to release was solved by two means. The apparatus was enclosed in a glass box, to shield it from air currents. Further, to the end of one of the sticks was attached a fine "whisker," which passed through a hanging drop of oil (Fig. 1D). Surface tension held the whisker centered in the drop. On release of the sticks the whisker pulled free, and because it was centered in the drop, little transverse momentum was imparted.

To cause the sticks to rotate to the vertical when released, they were pivoted a little off center so gravity would do the work. To hold them until release, Victor used a "latch" made of bimetal strip, which he triggered by the heat of a light beam sent in through the glass. A little "stickum" on the ends of the sticks made them stop dead at the vertical.

Victor said he used the experiment with success in the lectures at Caltech. Since the device is a torsion pendulum, he found that the best way of displaying the effect was by the ballistic throw—the maximum excursion of the light spot on the scale. With the device on the lecture table and the scale on the far wall, the excursion was about a meter—a striking effect. When Victor went to Thule, Greenland (for another purpose) he took the experiment along. He was interested because there the vertical component of angular velocity of the Earth is 14.6 deg/hr, as against 8.4 deg/hr in Pasadena. His result gave the right ratio to within 4 percent, but both figures were 20 percent high. He did not find out why.

The better part of a century ago, A.H. Compton, whose name is known for the Compton Effect, did a successful experiment[7] employing the conservation of angular momentum. He used a hoop-shaped glass tube, the order of a meter in diameter, filled with water. He supported it on a horizontal axle, on its diameter (Fig. 1E). The water in the tube contained some suspended matter, to make its movement visible. A microscope, which could be focused on particles of the suspended matter, was attached to the tube. The hoop was oriented so that the Earth's axis was normal to its plane, and enough time was allowed for the water to come to rest with respect to the tube, as indicated by the suspended matter. Both were then rotating with the Earth.

Next the hoop was flipped 180 deg around the axle, back into the same plane. The hoop and microscope were still rotating with the Earth, in the same sense. But the sense of rotation of the water was reversed. The counterrotation of the water persisted for only a short time, but it was seen and it was in the right sense. Ideally, its initial circulation relative to the microscope would be 30 deg/hr. That works out to over 4mm/min at the radius of the hoop, easily seen in the microscope.

Finally I'll mention two methods of sensing the rotation of the Earth that I

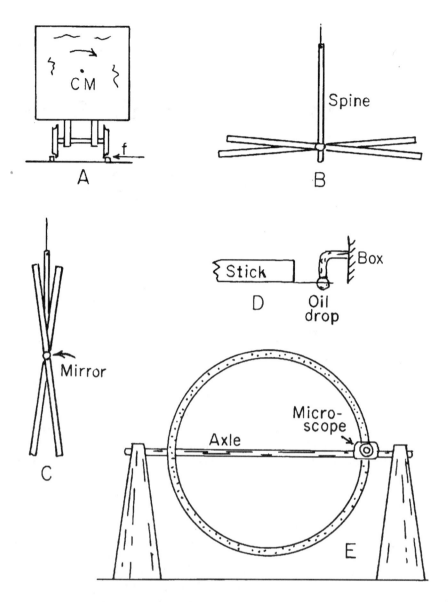

Fig. 1. A: An indication of how a horizontal force on the wheels of a car makes it tend to roll over. B: The two sticks in the position in which they are allowed to come to rest. C: The sticks, after rotating to the vertical. (The sketches show the sticks not quite parallel, so details can be seen.) D: The whisker in the oil drop, for damping residual rotation of the sticks prior to release. E: Compton's circular glass tube full of fluid, which can quickly be rotated 180 deg around the axle.

have already written about in "How Things Work." One[8] combines a very old method with very recent technology. It is the ring interferometer, dating from 1913, now increased in sensitivity by many orders of magnitude by fiber optics technology. It probably is the most precise of all the methods mentioned here.

The second[9] is the old story of water going down the drain in the washbasin—and supposedly going into rotation in opposite senses in the two hemispheres. As the water moves inward to the drain, its angular velocity should grow to many times that of the Earth. I am told that Ascher Shapiro's fluid dynamics group at M.I.T. demonstrated the phenomenon 30 years ago. It takes very skillful technique to detect the effect and remains a challenge to anyone who wants to duplicate the experiment.

References

1. Mark M. Payne, "Does the earth rotate?" *Phys. Teach.* **25**, 87 (1987).
2. H. Richard Crane, "The Foucault pendulum as a murder weapon and a physicist's delight," *Phys. Teach.* **28**, 264 (1990).
3. Robley C. Williams, Professor Emeritus, University of California, Berkeley, CA.
4. For an example of a demonstration, see Lester Evans, *Phys. Teach.* **20**, 102 (1982). For the general analysis of the motion of objects in a rotating system, see "Coriolis effect" in any theoretical mechanics book.
5. Here we have to consider a component of the Earth's angular velocity. Locally, the angular velocity of the surface with respect to an axis normal to it is the angular velocity of the Earth about its axis times the sine of the latitude. That formula will be recognized as the one used to predict the rate of turning of a Foucault pendulum.
6. H. Victor Neher, Professor Emeritus, Caltech, Pasadena, CA; Private correspondence, 1979 to present.
7. A. H. Compton, "The water tube gyro," *Science* **37**, 803 (1913).
8. "How Things Work," *Phys. Teach.* **24**, 52 (1986).
9. "How Things Work," *Phys. Teach.* **25**, 516 (1987).

Chattering, the Chatterring, and the Hula Hoop

I was given a toy[1] that I found to be interesting as a physics puzzle, although exasperating until I learned the trick of starting it.

The toy (Fig. 1), called a Chatterring, is a 28-cm-diameter hoop made from a ¼-in (6.4-mm) metal rod, on which there are five rings that have holes nearly twice the diameter of the rod. When given the right start, they execute spinning and wobbling motion, too fast for the eye to follow. (The photographs used here were taken by flash.) More shots are shown in the first two parts of Fig. 2. As the rings spin they progress downward, getting energy from gravity to keep spinning. Their downward speed is such that they would go the distance of the circumference of the hoop in about two seconds. But the game is to not let them go downward; the game is to rotate the hoop so as to just cancel their downward progress. Then they stay spinning at the same height until the player gets tired. Of course in this case the player supplies the energy for the spinning.

The name of the toy suggests that the rings chatter. But the chattering is not what groups of people or sparrows or monkeys do. The reference is to what happens in machinery when a shaft is rotating in a bearing that is badly worn. The shaft may suddenly change its mode of motion. Instead of sliding in the bearing, as it should, the shaft may *roll* around on the inner surface of the bearing, as in Fig. 3A. The solid arrow shows the sense of rotation of the shaft and the small numbers show where the shaft and bearing are going to make contact as the rolling goes on. The center axis of the shaft goes around as indicated by the dashed circle and the dashed arrow. The latter is what makes the machine vibrate. The frequency of rotation of the shaft and of its axis going around are not the same: the ratio de-

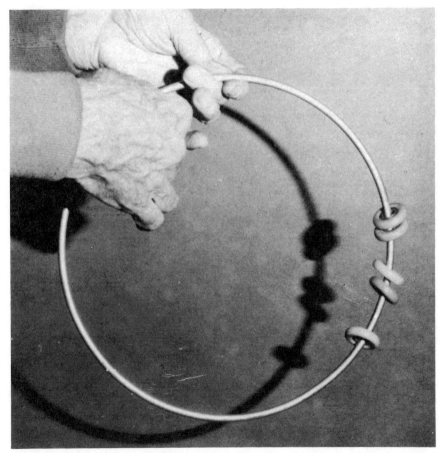

Fig. 1. Chatterring being rotated counterclockwise to keep the spinning rings at constant height.

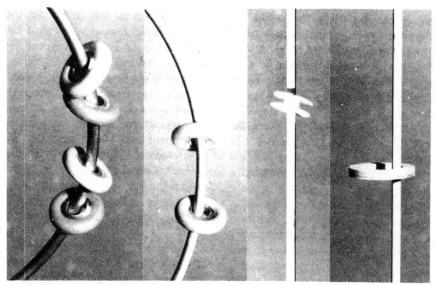

Fig. 2. Close-up flash photos of the spinning rings on the Chatterring, and two of the rings spinning on a dowel.

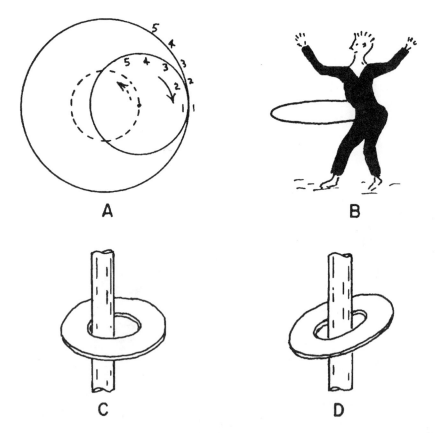

Fig. 3. A) A shaft rolling around the inside of a worn bearing. Numbers show places contact will be made as it rolls. Dashed circle is the path of the center axis of the shaft. B) The Hula Hoop. C and D) Sketches of a washer spinning on the dowel, as seen by the strobe effect.

pends on the ratio of diameters of the shaft and of the worn bearing.

So much for chatter in machinery. If you imagine it in reverse, i.e., the bearing rolling around the shaft, you have motion that is similar to that seen in the Chatterring. And if you remember far enough back, you've seen (or made) that motion with a Hula Hoop (Fig. 3B). But neither the worn bearing nor the Hula Hoop helps with the key question: Why does the point of contact of the ring with the rod trace a downward helix of such pitch that the ring gains energy at a rate that maintains the spinning?

To learn anything further one needs to be able to experiment by varying some of the parameters, especially the size of the hole in the ring. The Chatterring is not a good subject for experiment; its rings are of fixed dimensions, and they cannot be exchanged for others without cutting the hoop, thereby destroying the toy. But there is an easy way out. Needed are only a wooden dowel or other rod and a selection of rings or washers from the scrap box. I used a dowel of ¼-in (6.4-mm) diameter. It was vertical, held at the bottom in a vise, and prevented from moving at the top by the hand. The second two flash photographs in Fig. 2 show a plastic spool and a wooden ring spinning on the dowel. The spool moves downward at uniform speed, spinning. The large ring stays at the same height until it loses its spinning energy and then it drops.

Rings with other size holes were tried. Three facts were learned. First, to give continuous spinning, the ratio of the diameter of the hole to that of the rod has to be in a limited range. Ratios from 1.4 to 2.3 worked; larger or smaller did not. Second, the greater that ratio, the greater the downward speed. Third, all of the rings (in the range of hole-diameter that worked) maintained a slope with respect to the rod, at nearly as great an angle as permitted by the size of the hole.

Since the spinning was a blur to the eye, this way of observing seemed to be the end of the line—until a beautiful strobe effect showed up, by serendipity. It happened while I was trying metal washers. Washers are thin disks, one or two millimeters in thickness, with holes of various diameters. One I tried was a shiny one. The direction of illumination and the position of the eye happened to be such that a flash of specular reflection was seen at the same place in each revolution. That froze (to the eye) the rotation, leaving only the downward progression. The exact orientation of the washer with respect to the rod could be seen. A strobe lamp could not have done as well: this was even self-synchronizing. Try it! It takes only a shiny washer and a rod.

Figures 3C and 3D are sketches of the strobe image at two different phases of the rotation, as seen by locating the lamp and the eye at different positions. They are enough to show that the washer should progress down the rod. In the first, the washer slopes exactly downward toward the viewer. But the contact of the hole with the rod is not at its highest point, as might be expected. It is counterclockwise from that. There, the washer has a downward slope, so as it goes around and around, the point of contact will trace a helix down the rod. In the second, the contact with the rod is exactly behind the rod, as seen by the viewer. But the washer slopes in such a way that, again, the point of contact will progress downward in a helix on the rod.

Trying to see in a more analytic way why a ring assumes the attitude it does with respect to the rod defeats me. All the forces in the book are there: gravitational, centrifugal, frictional, and gyroscopic. The latter effect evidently seemed to the inventor to be important: the subtitle of the toy on the card is "gyro-walkers." Even if the toy is hard to understand, it is fun, especially because the friend you show it to will not be able to start the spin—as you can do so easily!

Reference

1. The Chatterring. Imported and sold by Fascinations Toys and Gifts Inc., 18964 Des Moines Way, Seattle, WA 98148; 1-800-544-0810.

Never Throw a 2-inch Speaker Away: It Has Many Uses

Recently I bought a pair of intercom units to replace an ancient pair that connected between my workshop and upstairs. These are the simple and cheap kind, linked by wires, not radio. Each contains a small PM (permanent magnet) loudspeaker. The interesting point, to me, is the simple way in which each loudspeaker serves either as microphone or speaker, to suit the direction of the talk.

First let's see how a PM speaker can play a dual role. Though the construction may be familiar to many readers, it is shown as Fig. 1A. Current in the coil, in the magnetic field, makes a force that moves the paper cone and makes sound. If, conversely, sound makes the cone move, the coil "cuts" magnetic field and generates voltage that can be amplified.

In the intercom pair there is an amplifier in one of the units—the "base" unit. Let's say the speaker in that unit is connected to the input of the amplifier. The output goes by wire to the speaker of the remote unit. So the person at the base unit can talk. To reverse the direction of talk—and reverse the roles of the speakers—input and output of the amplifier have to be interchanged (by a "push to talk" switch). In effect, the amplifier is turned end for end. As simple as that.

If you should experiment with a small speaker[1] as a microphone, you should use an amplifier that has microphone input, not one intended for input from a radio or tape player. The latter kind does not have enough gain. If you use a PA (public address) amplifier, watch out: its power could fry the speaker, which is rated at only a fraction of a watt. I use an inexpensive 1-W amplifier.[2]

The small speaker is easily modified into an electro-mechanical transducer, simply by cementing a cone of cork to its center (Fig. 1A). When the cork is

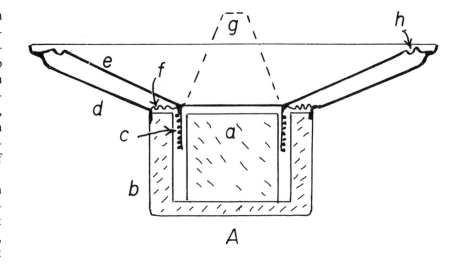

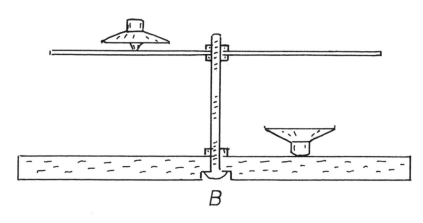

Fig. 1A. Anatomy of a loudspeaker, in cross section. a—cylindrical permanent magnet; b—soft iron cup; c—the coil, a single layer of fine wire on a thin plastic tube; d—conical metal support for the paper cone; e—the paper cone that moves and makes the sound; f—flexible seal against dust; g—dashed outline of a cork showing where it can be cemented to the center part of the paper cone; h—corrugation in the paper cone, to allow it to move in and out. Fig. 1B. The Chladni plate, supported on a bolt from a base of plywood. The driver speaker with the cork is pressed lightly on top and the microphone speaker is underneath. The support must be rigid: nuts tight.

gently pressed against something that is vibrating, the coil of the speaker generates a voltage—an electrical signal. Conversely, if current is sent through the coil, and the cork is pressed against some object, the object will be vibrated. It's a transducer in both directions.

The first of the above applications

was demonstrated in an earlier "How Things Work" column.[3] When the cork was pressed against a certain part of the hand and the muscles were tensed, a fluctuating voltage was generated by the coil. It was displayed on an oscilloscope and it showed that in tensed muscles there is a mechanical pulsation at

Fig. 2. Vibration patterns of the Chladni plate when the cork is in contact at six different places. For the bottom two, the plate is loaded at its edge with a small C-clamp, which can be seen.

near the surface. That makes a feedback loop: vibration of the surface makes sound that is picked up by the microphone and amplified to produce motion of the cork, which drives the vibration. The vibration may occur at any one of the several natural "ringing" frequencies of the object, depending on where the cork is making contact. There is one condition: the signal picked up by the microphone has to arrive at the driver in the right phase to drive the vibration. So it may be necessary to move the microphone around to find a location such that the phase is right.

The above feedback loop of microphone and driver works beautifully in making a Chladni plate[6] vibrate. The cork is pressed against the top side of the plate, and the microphone is placed underneath. A number of places of contact of the cork on the plate will produce strong vibration, each with a different pattern and a different frequency. As in the standard demonstration, sand sprinkled on the plate will dance and come to rest at the nodes, the lines of no vibration. Six patterns made that way are shown in Fig. 2. The plate was of ⅛-in (3.2-mm) aluminum, 26 cm square, supported on a bolt at its center as in Fig. 1B. For two of the patterns shown, extra mass in the form of a small C-clamp was introduced at the edge of the plate, as seen in the bottom two photographs of Fig. 2.

A word about electrostatic charge. Sand grains, or the plate if it is painted, can acquire static charge. That prevents the grains from dancing. For the patterns shown, the plate was cleaned with fine emery cloth. The sand used was magnetite, gathered from ordinary sand with a magnet. Magnetite is a poor insulator and besides it is black, so it shows up on aluminum.

References

1. Two-inch speakers from Radio Shack, if not from junk box.
2. Radio Shack "Mini Amplifier Speaker" with internal speaker bypassed.
3. *Phys. Teach.* **27**, 687 (1989).
4. *Phys. Teach.* **29**, 142 (1991).
5. "*Q*" (quality factor) is proportional (inversely) to the fractional loss of energy

10 to 20 Hz. In another "How Things Work" column,[4] the small speaker (without the paper cone) was shown to be the heart of the electronic weighing balance. In that, current is sent through the coil to produce a force that balances the weight of the object to be weighed. The current required to achieve balance tells the weight of the object.

There is an interesting and simple application of the speaker with the cork,

if it is used as a driver rather than as a sensor. The plain speaker (acting as a microphone) is connected to the input of an amplifier, and the one with the cork to the output. Anything that is capable of vibrating with a high Q,[5] say a stainless steel cooking pot or a stem water glass, can be made to sing loudly. To make that happen the cork is pressed lightly against the surface, and the speaker acting as microphone is held of vibration, per cycle, of a freely vibrating object such as a bell after it has been struck. For a given bell, the higher the Q value the slower the vibration would decrease. Q is used for electrical circuits, as well as mechanical vibrators. See *The Physics Teacher* centerfold, December 1990.

6. A demonstration originated by Ernst F.F. Chladni in 1781. Since then it has been standard in the repertoire of physicists. Most often the plate is made to vibrate by bowing its edge with a resin-coated bow.

Magnetic Levitation (Almost)

A friend was given a toy called *Revolution*,[1] purported to be frictionless and to display perpetual motion. Of course I was intrigued so I borrowed it. Briefly, it is a spinner made with two disk-shaped magnets on a wooden axle, supported by the repulsion of four rectangular magnets that are hidden in a wood base. Figure 1 is a photograph and Fig. 2 is a diagram, looking from above, of the arrangement of the magnets, with their polarities indicated. All are magnetized parallel to the axis of the spinner. Transversely, the spinner is stable, lying in a potential trough. But axially it is at the top of a potential hill and therefore unstable. If released it would escape axially in either direction, with acceleration.

To stabilize the spinner axially, it is allowed to reside slightly off the top of the hill. The resulting force is opposed by a needle point on the end of the spinner, which rests on a glass disk on the holder. (The glass is a mirror, for no particular reason.)

The gadget invites experimentation. The literature says it will spin for 20 min. That was easily tested. It was true, approximately, starting from about 10 rps. A more detailed timing of the slow-down from 10 rps gave the following: At the end of 5 min, 5 rps; 10 min, 2 rps; 15 min, 1 rps; 20 min, rotation changed to oscillation over only part of a revolution, due to slight unbalance; 25 min, dead.

We wonder what dissipates the energy at such a slow rate. Three ways come to mind. Air friction, friction of the needle point, and eddy currents (currents induced by changing magnetic flux in a conductor). The last-named can be dismissed, because all the magnets are nonconductors (ceramic) and no metals are present. Air friction should decrease rapidly with the decrease in rate of rotation. The friction of the needle point should be about constant, since it is essentially sliding friction. A way to check on the air friction (e.g., by running the device in a vacuum) was not available. There is, however, a simple way to check on the friction of the needle point.

The force of the needle against the glass, and therefore its friction, is easily changed by tilting the whole device. First a run was made with the end of the base near the needle raised until the spinner was almost unstable axially, then a run with the opposite end raised the same amount. The times, from 1 rps to the end of the rotation, differed by not more than a minute. That showed, to my surprise, that even at less than 1 rps, the needle friction contributes in a minor way to the slowing. Maybe a reader who has a vacuum chamber will check the air friction.

Magnetic levitation is in the news these days, mainly stimulated by the advent of high-temperature (liquid nitrogen temperature) superconductivity. Experiments and plans are under way in several countries for levitated trains, in most cases employing superconductiv-

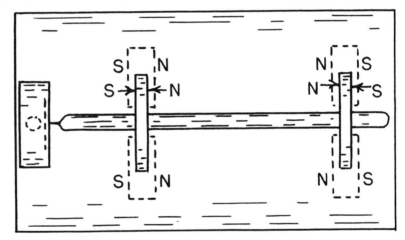

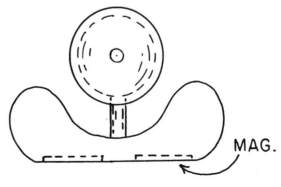

Fig. 1. "Revolution," toy showing the spinner supported in a magnetic field.

Fig. 2. Scale drawings of the toy, top and end views, showing especially the polarities of the magnets on the spindle and of those in the under side of the base (dashed lines).

Fig. 3. The author, levitated, taking the photograph you see, via two mirrors.

ity.[2,3] Levitation of a permanent magnet over a high-temperature superconduc-tor has become a simple and fascinating demonstration,[4] using kits of materials available from apparatus supply companies.[5]

Without superconductivity, stable levitation of a permanent magnet over an arrangement of other permanent magnets (such as the arrangement in the toy described), is another matter. There is a theorem by Earnshaw,[2] dating from long ago, that applies to any attempt to levitate a magnet by an array of other magnets. It says that if there is stability in one direction there must be instability in another direction, just as is evident in our toy.

Levitation of a reverse kind, a magnet or a piece of iron suspended below a magnet by attractive force, is possible and is an effective demonstration. The suspended piece (which may be concealed in the head of a doll) will spin, damped only by the air. The catch is that for vertical stability the upper magnet must be an electromagnet whose current is controlled by feedback. The standard method is to have a light source on one side and a light sensor on the other, so that if the suspended piece rises too high it cuts off some light, which brings about a decrease in current in the magnet coil, and vice versa.

Levitation is readily accomplished by mirrors. Figure 3 shows the author taking the photograph you see, via two mirrors. Can you figure the arrangement of the mirrors? The mirror system, called "levitator" is a popular exhibit in the Ann Arbor Hands-On Museum.

References

1. Listed in various novelty catalogs, e.g., The Nature Co., P.O. Box 188, Florence, KY 41022; "Perpetual Motion," $28.
2. T.D. Rossing and J.R. Hull, "Magnetic levitation," *Phys. Teach.* **29**, 552 (1991). This includes a discussion of the Meissner effect and of Earnshaw's theorem.
3. G. Stux, "Air trains," *Sci. Am.* **267**, 102 (1992).
4. P.J. Ouseph, "Levitation of a magnet over a superconductor," *Phys. Teach.* **28**, 205 (1990).
5. Edmund Scientific Co., 101 E Gloucester Pike, Barrington, NJ 08007.

Simple Devices that Fascinate—But Only Some People

Some of us can't resist simple-minded devices, especially the ones that serve no useful purpose. I'm one; I hope you are another. I'll continue where I left off in the column of December 1991.

At least 20 years ago I visited a woodworkers' fair in Gatlinburg, Tennessee. One exhibit stuck in my mind. I had to reproduce it (Fig. 1) so as to work it and understand it. The most interesting point to me is that it divides or multiplies by two, without the use of gears. So it's a binary counter—it could be the basic element for an adding machine.

The device consists of a wooden disk having two tracks at right angles, and supported from below on an axle, so it can be rotated by hand. In the tracks there are sliders, linked together on an arm. If the arm is held pointing in a fixed direction and the disk is rotated, the

Fig. 2. A mechanical amplifier: 1 kg of lead is being raised by the little finger.

sliders move, not quite colliding. The disk has to turn two revolutions to make the arm go back and forth once. Division by two. Of course it multiplies by two if the arm is moved back and forth. It would not make a very useful computer; its better use is on the desk top, to make people ask "What the dickens is that?"

This year, out of the blue, my friend Betty Wood[1] sent me a sketch of the same device, but with an application new to me. She pointed out that if you attach a pencil to the end of the arm, hold the disk fixed, and move the arm around, you have a machine for making perfect ellipses. She supplied the calculation to prove it, done by Rae Young. One can think of applications using a clock motor to keep it turning, say for a clock with an elliptical face.

The things that can be done with pulleys and strings are without end. They go beyond the usual block and tackle; for instance a force or torque amplifier. Such devices have been used in instruments, as well as for moving heavy loads. They depend on a motor for power and on the clever use of friction. Figure 2 is a force amplifier. A turn of the string is wound around a drum on a motor shaft, which is rotating all the time. A small pull on the string tightens it on the drum, and the motor does the work of raising the load. It's interesting that the tightness of the string on the drum is self-adjusting; that is, the load will be raised at any rate the string is pulled, or will be stable at any height if the pull on the string pauses. The figure shows a 1-kg lead block being raised by the little finger. The motor is a geared-down one, with shaft speed 1 rev/s, and there are 1½ turns of string on the 1.3-cm-diameter wooden cylinder.

The Chinese windlass, invented in the far distant past, is interesting for two reasons. Its mechanical advantage can be made anything up to infinity, and the load being raised will stay at any height without force on the crank handle. String winds onto one cylinder while it winds off the other. In the model shown (Fig. 3), the difference in circumferences is 1 cm. Two strands of string support the load, so for one turn the load goes up or down a half centimeter. The crank handle goes 34 cm, so the mechanical advantage is 68. The load does not go down when there is no force on the handle because the energy produced in a ½-cm fall is not enough to turn the cylinders one revolution against friction. That of course would not be true

Fig. 1. A device that divides and multiplies by two, and also generates ellipses.

Fig. 3. The Chinese windlass. String winds and unwinds on two cylinders of nearly the same radius. Possible mechanical advantage is unlimited.

if the cylinders were more different in radius.

We now come to a mechanism that seems to go the wrong way when a string is pulled (Fig. 4). It is the basis of the old toy, the Climbing Monkey.[2] Two

Fig. 4. A device that seems to go the wrong way when the lower string is pulled. It is the basis for the Climbing Monkey toy.

spools of different radii, each containing many turns of string, are (in the model shown) glued together. The

upper string is tied to something above, so that the device hangs. If the lower string is pulled, string unwinds from the spool of larger radius and winds up on the one of smaller radius. Because of the difference in radii, a given force applied to the lower string produces a greater force in the upper string. That makes the whole device go up.

In our model there is a pulley below the double spool. It is free-turning, and its only function is to guide the lower string so it will come out at the center of the device.

The world is full of gizmos, only some of which display interesting physics. Readers are invited to suggest candidates for a future column.

References
1. Dr. Elizabeth Wood, retired from Bell Telephone Laboratories.
2. Henry Levinstein, "The physics of toys," *Phys. Teach.* 30, 358 (1982).

The Frictionless Bird and the Aerodynamic Tortoise

A statement in a review of a book[1] caused a yellow flag to go up in my mind. To quote: "Flying, like other physical activities, costs energy, and a frictionless bird, having attained level flight and satisfied with its course, could glide forever, without moving a muscle." My reaction was, "Okay, so he doesn't have friction (drag), but how does he stay up?" One has to look to see where the upward force, equal to his weight, comes from.

To keep from falling, a bird has, continually, to push downward on the air. Same reason that when standing we have to push downward on the floor with a force equal to our weight. But the difference is that the floor doesn't yield

and the air does. By pushing downward on the air, the bird leaves a trail of air behind that has a downward velocity. Which means that he leaves energy behind, energy that he has to supply.

The *frictionless* bird would only have to push the air downward, to stay aloft. A *real* bird has to give the air some component of velocity backward as well, to keep himself going forward against air friction.

It's interesting to pursue the consequences of pushing air downward. What fills the space left by the air that is pushed down? And where does the air go that is pushed down? The answer is the only logical one: the air pushed down circulates up and around to fill the

space it came from (Fig. 1A). The circulation is symmetrical on the two sides of the bird, so no net angular momentum is given or taken. The circulation persists for only a short distance behind the bird before its energy is converted to heat. And that is how the bird's output of effort ends up. Circulation is a term much used in aeronautics; the pattern is the same for an airplane as for a bird.

To pursue still further: many kinds of large birds are seen to fly in a formation. Each bird, by flying to the side of the bird ahead can avoid the downdraft of air (Fig. 1B). Also, by flying a little behind the bird ahead he may get a little free lift from the upward movement of the air in the circulation pattern. These conditions result in a V formation (Fig. 1C).

Birds don't calculate, but apparently they know how far back and to the side to position themselves to have the easiest flying. So it is not surprising that the V, say for geese, has about the same angle at its vertex, whenever seen. It could be different for birds of different sizes and speeds. One might guess *n* geese in V formation require a total power less than *n* times that required by a lone goose. Otherwise why would geese fly in a V?

From the aerodynamics of a goose to that of a tortoise is a big jump. On the cover of an issue of *Snippets*[2] there is a cartoon (by permission of *Punch*) of two tortoises, one saying to the other, "It's infuriating. Aerodynamically speaking, we're capable of two hundred miles an hour." Inside the same issue there is a discussion, presumably by the editor, addressed to "serious thinkers." A sketch of the tortoise is shown with the lines of air flow over her curved shell and under it. I've drawn a similar diagram (Fig. 2). From the different distances the air has to travel, assuming the same time of travel, the two velocities are found, and (by the Bernoulli equation[3]) the lift. The conclusion reached is that the tortoise would lift off the ground at 80 mi/hr (30 m/s).

The matter was not allowed to rest there. In a later issue[4] none other than our erstwhile President of AAPT, Albert Bartlett,[5] makes two points. He sees no reason to assume that the air travels above and below the tortoise in the same length of time; therefore that the assumption about the two velocities has no basis. Second, he prefers to use Newton's laws instead of the Bernoulli equation to see whether there is lift or not. He makes the point (as we did here in connection with the bird) that if there is to be lift, the air behind must be left with a downward velocity. He leaves it there, without comment on the fact that the figure in *Snippets*, as in our Fig. 2, does not show the air having a downward velocity behind the tortoise. Would he (or you) say there is no lift?

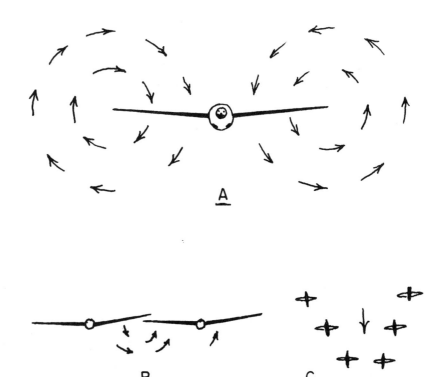

A

B   C

Fig. 1A. Circulation of the air produced and left behind by a bird or plane when gliding. B. Bird on the right flying to the side and behind the bird ahead. C. V formation that birds evidently find to provide the easiest flying.

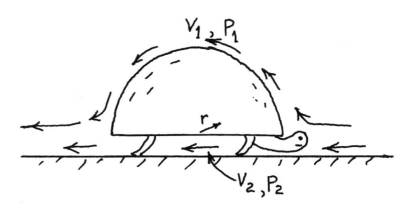

Fig. 2. Pattern of air movement relative to a moving tortoise, drawn following the illustration in *Snippets*.

I decline to take sides. The problem certainly has been solved. It is important to automobile designers. While they might not like to have it pointed out, the profiles, above and below, of tortoises and automobiles (especially the Volkswagen "bug") are nearly the same, as far as aerodynamics goes.

The *Snippets* editor makes a parting comment: "When I started with the tortoise in *Snippets*-1, I never expected to raise so many *hares*!"

References
1. Alan P. Lightman, *Time and Travel and Papa Joe's Pipe* (Scribners, New York, 1992), reviewed by Victor Wilson in Newhouse News Service.
2. *Snippets*, No. 1, May 1982. (*Snippets* is a newsletter on physics education, published by The Institute of Physics, London.)

3. The Bernoulli equation, $p_1/d + \frac{1}{2}v_1{}^2 = p_2/d + \frac{1}{2}v_2{}^2$ is of course consistent with Newton's laws; it's another way of getting the lift. As applied to the tortoise, v_1, p_1 and v_2, p_2 are the velocities and the pressures in the air above and below; d is the density. *Snippets*' editor assumes that the times of travel of the air above and below are the same and that the tortoise's shell is a hemisphere. That gives $v_1 = \frac{\pi}{2} v_2$, and from that, $p_2 - p_1$ in terms of the speed of the tortoise. A little summing over the area gets the lift.

4. *Snippets*, No. 5, winter 1983–84.

5. Department of Physics, University of Colorado, Boulder, CO 80309-0390.

The Quartz Analog Watch: A Wonder Machine

People live closer to quartz wristwatches than to any other kind of equipment, yet few (even among physicists) would be able to describe in more than general terms how they work. That's partly because the watches are so small and compact that even when the back is opened (Fig. 1), the works are no less a mystery. It took a number of inquiries to watch companies to get explanations that satisfied me and that I thought would satisfy other physicists.[1] Quartz watches come in two categories: analog, the ones with hands, and digital, the ones with the liquid crystal readout. One kind, the analog, will be enough for this time around. Digital may be left for later.

A word here about the term "quartz watch." The term is short for any watch whose time reading is derived from the vibration of a crystal through the *piezoelectric effect*,[2] the standard crystal for watches being quartz. An electronic circuit drives the crystal in vibration at its resonant frequency. The resonant frequency is extremely high. Stepping the high frequency down and using it to control the turning of the hands (of an analog watch) is the job of 90% of the parts found in the watch.

The description falls into three parts: (1) the quartz crystal, its form, and how its frequency is set to a standard; (2) the integrated circuit (IC) chip that drives the crystal in vibration, scales its frequency down, and forms pulses that turn the motor; and (3) the motor that drives the train of gears that turns the hands. The description here will be of the watch of a particular company; other

Fig. 1. Quartz analog watch with back cover removed. **a**, capsule that contains the quartz tuning fork; **b**, coil that momentarily magnetizes the stator of the motor; **c**, axle of the motor armature; **d**, integrated circuit (IC); **e**, gear box; **f**, concentric axles of the three hands; **g**, battery.

makes may differ in details, but not in basic principle.

When I learned that the resonant frequency of the quartz crystal in a watch is only a little over 30 *kilo*hertz, I was greatly puzzled. All the crystals I had worked with in my ham radio days were in the *mega*hertz range. And the smaller the crystal the higher was the frequency, so how could the frequency of the al-most microscopic crystal in a watch be so low? The answer, I found, is that the crystal for the watch does not vibrate in the standard mode, that in which opposite faces move toward and away from each other. The crystal is cut in the form of a miniature tuning fork,[3] the two arms of which move toward and away from each other (Fig. 2A). The tuning fork is mounted in an evacuated metal capsule, a in Fig. 1, whose outside diameter is only 1 mm. I have not opened one of the capsules, but can imagine how small the tuning fork must be.

The second marvel about the tuning fork is how its frequency is adjusted *on a production line* with such accuracy that the watch will meet the company's claim of losing or gaining no more than a minute a year. That translates into two parts per million in frequency. Adjustment can only be done physically: by changing the shape or distribution of mass. Now enter the laser. The frequency of the tuning fork as cut from the crystal of quartz is, intentionally, a little too high. Then spots of gold are deposited (probably by evaporation) near the ends of the arms, to add enough mass to make the frequency a little too *low*. On the production line, while the frequency of the tuning fork is being compared (electronically) with the standard, a laser beam starts evaporating the gold of the spots, causing the frequency to increase. When the frequency matches that of the standard, the laser beam cuts off, and is ready for the next tuning fork.

To see what the IC chip has to do, we can start with the fact that the second hand of the watch (and the other two hands also) advance discontinuously once every second. The IC chip scales frequency down by successive divisions by 2. Therefore the turning fork frequency in Hz must be a power of 2; in fact it is the 15th power, 32,768 Hz. One minute per year for the watch translates into matching the tuning fork frequency to that of the standard to within 0.06 Hz. Evidently that is doable, in production-line fashion.[4]

It's interesting to think of the overall reduction of the frequency of the tuning fork. To the rotation of the hour hand it is 2 to the 21st power, and if the watch has a calendar, it is 2 to the 26th!

We come to the last of the three ingenious parts of the watch: the motor. It is one of a class called a stepping motor. The name describes it: in response to an electric pulse, the rotating part (the armature) advances by a finite angle and stops, to await the next pulse. The stepping motor, diagrammed in Fig. 2B, can

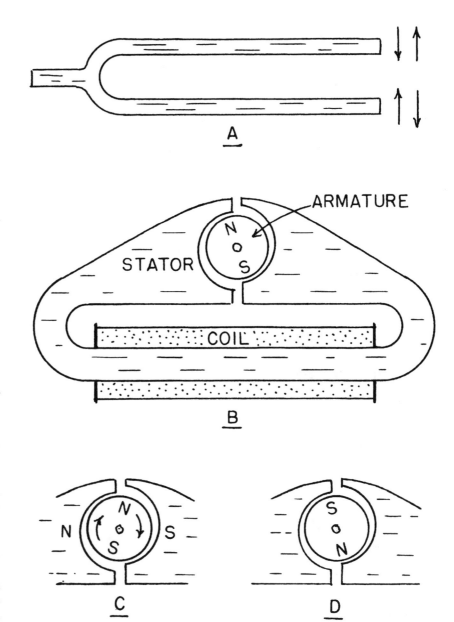

Fig. 2. **A**, a conventional tuning fork; vibrates the same way as the one in the watch. **B**, the motor, with the armature in the resting position when the stator is unmagnetized. **C**, current in the coil, stator magnetized in the sense shown, armature starting rotation, clockwise. **D**, stator again unmagnetized, armature comes to rest as shown, 180 degrees from starting position.

be identified in Fig. 1 by its coil, b. The soft iron stator is concealed by a plate that holds it in place. The axle of the armature is at c.

In Fig. 2, B, C, and D show the action. The soft iron stator, normally nonmagnetic, can be made magnetic momentarily by current in the coil. The armature is a permanent magnet. The key to the stepping action is that the armature is not in a cylindrical cavity in the stator.

The halves of the cavity are offset, as shown. When there is no current in the coil, the armature seeks a position such that its poles are nearest to the soft iron, as in both B and D. While it is residing at rest in that position, the coil receives a pulse of current from the IC (d in Fig. 1), making the stator magnetic, with such polarity that the poles of the armature are repelled. That makes the armature start rotating in the sense that will get its poles farther from the stator, namely clockwise, as in Fig. 2C. The current in the coil lasts only long enough to start the armature rotating. Once started, it goes on to find the next position in which its poles are nearest the (nonmagnetic) soft iron, which will be 180 degrees from where it started. The next pulse of current in the coil (which must be in the opposite sense to that of the first one) will start the armature on the second 180 degrees of clockwise rotation. So that is the response to the current pulses from the IC, which are short, alternating in direction, and spaced one second apart.

The rest of the story, from the stepping motor to the turning of the hands, is old stuff: a train of gears (e in Fig. 1) that reduces the rotation of the motor to the appropriate rate for each of the three hands, whose concentric axles are at f. Not to forget the battery, g, the biggest component of all.

References

1. We are indebted to Dan Fenwick, Swiss Watch Technical Center, 1817 William Penn Way, Lancaster, PA 17601 for a very informative technical pamphlet.
2. A more complete discussion of piezo-electric oscillations will be found in a footnote in an earlier "How Things Work" column [*Phys. Teach.* **26**, 120 (1988)].
3. The use of a tuning fork is not new. The Accutron, which had its day before the advent of quartz watches, had a steel tuning fork, of frequency 300 Hz, driven by a battery and one transistor. The rest was mechanical: first a ratchet wheel, then a gear train leading to the hands. The watch, held to the ear, was used by doctors to check hearing.
4. Another case of precision in mass production, described in these columns is how steel balls, spherical and uniform to 0.0001 inch are made, cheaply, by the billions [*Phys. Teach.* **24**, 561 (1986)].

A Reluctant Hourglass and a Backing-down Woodpecker

More than two decades ago I was given a puzzler, evidently handmade as a hobby by a glassblower who now cannot be traced. It will confound a physicist for at least a few minutes. I have not seen it in catalogs.

The device consists of an hourglass, in a sealed glass tube about 5 cm in diameter and 35 cm long filled with a colored liquid, probably water (Fig. 1, left). The hourglass has such overall density that it will just float. Starting from the stable condition with the hourglass floating and all the sand in its bottom bulb, you invert the tube, so the hourglass is inverted and at the bottom of the tube (Fig. 1, right). Surprise: the hourglass stays at the bottom for a minute or so, before beginning its rise to the top. Why the delay?

The explanation you would come to after cycling the gadget a few times is the following. After inverting the tube, the sand, being in the upper bulb, makes the hourglass top-heavy. So it tries to turn itself upside down. That causes its upper and lower bulbs to press against the walls of the tube. The resulting friction keeps it from rising, until more than half of the sand has run down, at which time it is no longer top-heavy. The diameter of

Fig. 1. The hourglass in the sealed tube filled with water. Left: floating at the top. Right: immediately after the tube is inverted. The hourglass is at the bottom, sand in the upper bulb, making it top-heavy.

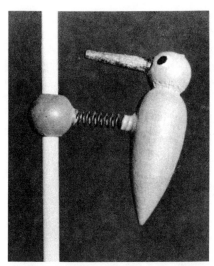

Fig. 2. The woodpecker on a stick. When vibrating (pecking) he will move downward.

the hourglass is almost as great as the inner diameter of the tube, so the slight inclination of the hourglass while it is top-heavy is not noticed. I tested a physicist-university president on it: it took three inversions.

The next puzzle has been around a long time: the woodpecker that backs down a stick, pecking as he goes. Betty Wood[1] sent me one, store-bought (Fig. 2). The construction is self-evident except for the important fact that the hole in the wooden ball is about 20 percent greater in diameter than the stick. Given a start by pulling the bird up or down (bending the coil spring) and releasing it, the bird backs down the stick at constant speed and with constant amplitude of the pecking motion. The bending of the spring and the motion of the bird are

in the vertical plane. The question put to me: How does the bird do it?

I found that not much could be learned by watching the bird perform; the pecking is at too high a frequency to follow. More could be learned by manipulating the spring and bird slowly by hand, with attention to what the oversize hole does with respect to the stick. By bending the spring upward and downward the amount that it bends during actual performance, it is seen that when it is bent downward the hole is oriented as A in Fig. 3, and when bent upward the hole is as B. So during vibration, the hole must switch, suddenly, from the one orientation to the other, at each instant the spring goes through neutral, changing from upward to downward bend or vice versa. So the problem is: How does that make the ball with the hole in it "walk" down the stick?

The real ball with the hole being too small and fast moving for the eye to follow, I turned to a coffee can, open at both ends, surrounding a 4-cm-diameter post. The top and bottom rims were labeled a, b, c, and d (Fig. 3C). The can could be taken through trial sequences of motion slowly by hand. It turned out there was a unique sequence that would make the can progress downward, *walking*, not sliding. It was, starting from C: pivot about d to get D; pivot about c to get E; pivot about b to get F; pivot about a to get back to C. That's a full cycle, but note that the can has ended up *lower* on the post. Compare where a is at the start and at the end.

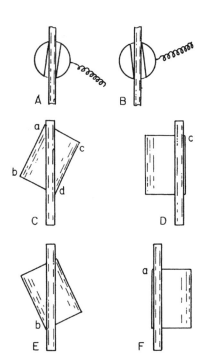

Fig. 3. Cutaway views: A and B, positions of the wooden ball on the stick, when the spring is bent downward, and upward; C to F, positions of the coffee can on the pole, as explained in the text.

There is an alternate sequence of movements that makes the can *climb* the rod. Why the bird prefers the "right" mode is not clear. A point about the energy: Of course the energy dissipated in air friction, hysteresis of the spring, etc. comes from the bird's movement downward under gravity. But tracing it in detail looks a little complicated.

Reference
1. Elizabeth Wood, Bell Laboratories (retired), has helped with several "How Things Work" articles.

How Antilock Brakes May Save You from a Spin

The TV ads for cars nowadays are full of talk about antilock brakes, as well as air bags (the latter of which have been described in one of these columns[1]). For your (and my) curiosity about the antilock braking system (called ABS) I was able to get several engineering bulletins as well as a personal letter of description from Dean Lewis of ITT Automotive.[2]

Everyone who has driven on icy, or even wet, pavement knows that when braking, once a skid begins, turning the steering wheel has little effect. The reason of course is that when sliding, the force of friction is against the direction of *sliding*, without regard to the orientation of the tire or the vehicle.

When braking on a slippery road, the optimum condition would be the application of braking just short of that which would initiate skidding. Then steering control would not be lost. But how is a driver (or a computer) to sense when enough is enough: when the threshold of skidding has almost but not quite been reached? He, she, or it cannot. The only way is to increase the braking until skidding starts, then quickly back away a little. A sampling process. The computer-controller in the ABS does just that, repeatedly. Thus it's a trade-off: the tire can be nonskidding only for part—at best most—of the time. But that's enough to leave the driver with steering control, and as much braking as practically possible.

Because the road condition under the tire may change rapidly, the sampling of skidding has to be done in rapid fire—up to 10 times a second. That causes a pulsing in the motion of the vehicle that can be felt by the passengers, and, as you may have noticed in the TV ads, discontinuous rotation of the wheels.

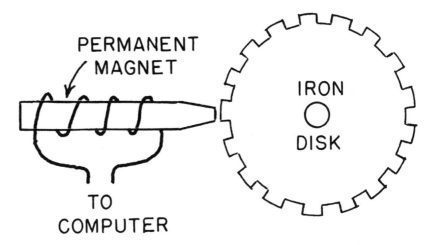

Fig. 1. Sensor for the rate of rotation of a wheel. As the teeth of the iron disk go by the permanent magnet, the flux linking the coil varies, producing electrical pulses for the computer. The pulse rate tells the rate of rotation.

The driver initiates the braking, and when in a tight situation will push the pedal all the way down. The job of the computer and control system then is to sense the onset of skidding, stop the increase in braking (in spite of the driver's heavy foot), and go into the sampling routine. The computer senses the start of skidding by a rapid change in the speed of rotation of the wheel; in the extreme, if the wheel locks, the speed drops to zero. The output of the computer controls the hydraulic pump and valve system that determines, from moment to moment, the intensity of braking.

How the onset of skidding is sensed is simple. On the car wheel there is fixed a toothed disk, like a thin gear. A permanent magnet with a coil wound around it is in position with its end close to the teeth (Fig. 1). The toothed disk is ferromagnetic, so as the teeth go by, electric pulses are generated, which go to the computer. The frequency of the pulses is a measure of the speed of rotation. It drops rapidly if skidding starts.

Braking is applied (brakes are tightened) by fluid pressure in a "hydraulic" system. The system consists of pumps and valves, ending in actuating cylinders at the wheels. These valves and parts are controlled electromechanically on orders from the ABS computer. There are variations from one car make to another in the way the control is done, and even differences in the names of the components, none of which we need to know for our purpose here.

Our discussion so far has not dealt with the fact that cars have four wheels—two, or in some cases four, of them being driver wheels. That calls for ABS control on the wheels individually, at least on the ones that are the most critical. Critical ones are first those that involve steering: the front ones. In some front-wheel-drive cars the rear wheels are not controlled individually, but in some they are. It is of course more important when the drive is in the rear.

There are two further applications of the systems for controlling slipping or skidding. One is the reverse of ABS: it

---

automatically limits the power applied to the drive wheels, rather than controlling the braking. On a slippery road, if the driver applies too much engine power to the drive wheels, they may start to spin. Steering control is then reduced. The sensor on the wheel tells the computer that the rotation of the wheel has suddenly increased, indicating spinning. The remedy is less engine power, which the computer brings about by cutting down the "gas." From there the sample and control loop comes into play. The application is called automatic traction control.

The second application is a traction assist system. Those who live in ice-and-snow country know the hazard of parking with the right wheels in any icy gutter. When trying to start, the right drive wheel spins on the ice and the left one, although not slipping, is of little help in moving the car. That is the result of the differential gears, which are interposed between the right and left drive wheels on all cars (Fig. 2).

A reminder about the reason for the differential and what it does is in order. When a car is moving in a curve, the outside wheels have to rotate faster than the inside ones. The differential lets the wheels rotate differently; only the average of their rates being fixed by that of the drive shaft from the engine. A differential between the nondriving pair of wheels is not necessary: they can rotate independently. An exception is the four-wheel-drive car; it has a differential front and rear.

A result of the differential is that the tractions exerted (forward forces imparted to the car) by the right and left drive wheels are the same. So when the right one is spinning on ice, the traction exerted by the nonspinning one can be no more than that exerted by the one that

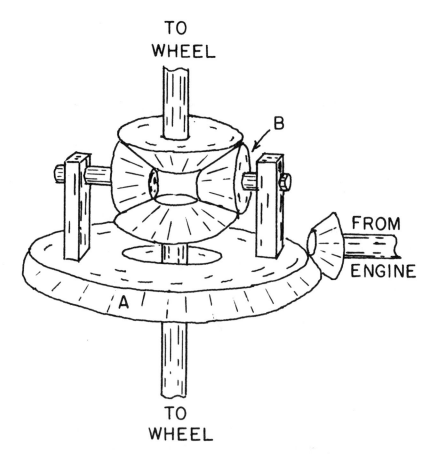

Fig. 2. The essential elements of a differential. When the car is going straight, the two car wheels rotate at the same rate; therefore the small gears B do not rotate. The rate of rotation of the car wheels and that of the large gear A are the same. When the car is turning, the wheels must rotate at different rates, and the gears B must rotate in opposite senses. The increase in rotation of one car wheel is the same as the decrease for the other, and their average is equal to that of gear A.

is spinning—and that's very little traction.

The solution, in the traction control system, is to apply enough braking to the wheel that is on ice to stop the spinning. Then the wheel on the dry surface will exert traction up to the point where *it* starts slipping—which is considerable traction, and away the car goes.

As a matter of interest to four-wheel-drive car owners, you too can get stuck in an icy gutter. Because there is a dif-

ferential at each end of your car, if the two right wheels spin on the ice, traction will be lacking for both of the left wheels.

References
1. H. R. Crane, "The air bag: An exercise in Newton's laws," *Phys. Teach.* **23**, 576 (Dec. 1985).
2. 3000 University Drive, Auburn Hills, MI 48321-7016.

The Quartz Watch with Digital Readout

A recent "How Things Work"[1] explained the quartz analog watch, with a promise of later attention to the kind with the liquid crystal digital readout. The packet Daniel Fenwick[2] sent me for the analog watch is an equally fertile source for the digital. So I am doubly indebted to him.

Digital watches are popular with people who do not have to fish for their glasses to see the numbers (which leaves me out). They cost less than analogs, and some of them do more things, even calculations. All due to the wonders of integrated circuit (IC) chips, a few mm square and costing but a few cents apiece.

To get started, I did "reverse engineering" to examine the parts and to get a photograph of their arrangement (Fig. 1). A colleague in the physics department, Fred Hendel, told me where to get digitals for 99 cents.[3] So there was little to lose by sacrificing one or two.

The first part of the digital watch is identical to that of the analog: a quartz tuning fork that is the master timekeeper, and a train of 15 divide-by-two's in the IC chip, which brings the tuning fork's frequency of 32,768 Hz down to 1 Hz. But from there on, instead of a motor and gears moving the hands of the analog, the chip in the digital keeps on reducing the frequency: in stages to give outputs for the display of the minute, hour, day, and month. The reduction, from 1 Hz down to the pulse that advances the month is about 21 more divisions by 2. "About 21" is said because the divisions, e.g., by 60 or 31, are not by even powers of 2. But not to worry; the circuit in the IC takes care of it. It even takes account of the 28, 30, and 31 days in different months. And all that is in a no-frills watch in the $5 range.

The other interesting part of the digital watch story is how the numbers are displayed. It's an ingenious use of polarization effects. If you look at the display through your polaroid glasses and

rotate the watch, the background of the display will change from light to dark, showing that the light that comes out of it is polarized. The numbers stay dark because they emit no light at all.

How the display works was described in these columns more than a decade ago.[4] But another look will be worthwhile, using current information. The display is a multilayer sandwich. A cut through it is seen in Fig. 2. It is less than 2 mm thick, and of surface size to suit the watch, as seen in Fig. 1. The important part of it is a layer of liquid crystal in the middle, between a pair of coated glass plates. It consists of long

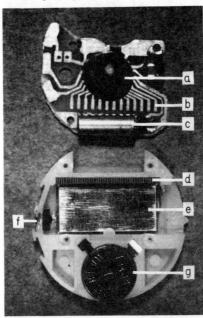

Fig. 1. The insides of a quartz digital watch, unfolded. a, integrated circuit chip (IC). b, electrical terminal fingers of the IC. c, capsule containing the quartz tuning fork. d, electrical terminals of the display, which include connections to all the segments of the numbers. e, display, seen from the rear. f, one of two switches, operated from push-buttons on the rim of the watch. By combinations of presses, the time can be set, or other functions can be shown, e.g., the seconds. g, battery. To assemble the two parts, as they are in the watch, move the top one downward until the tuning fork capsule fits into the notch, then close it like a book. The fingers of the IC (b) will make contact with the terminals of the display (d).

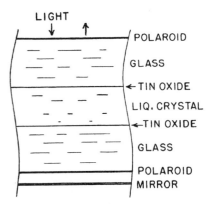

Fig. 2. A cut through the multilayer display unit, showing arrangement of the components. Functions of the components are explained in the text.

molecules, which, if not disturbed, will arrange themselves with their long axes parallel to one another, as a pseudocrystal.

The coatings (tin oxide) on the plates are striated: they have a "grain" of microscopic parallel scratches or grooves, achieved by rubbing or combing the coatings in one direction. When the liquid crystal "crystalizes" near and in contact with that surface, the molecules align with the grain. That would set the pattern for the alignment in the rest of the liquid crystal—*except for the fact* that the grain on the opposing coated plate is at right angles to that on the first. The liquid crystal accommodates itself to that, while remaining orderly, by *twisting* through the 90° between the grain direction of the one plate and the other (Fig. 3A). So it matches the grain top and bottom.

The Polaroid at the top of the stack is oriented so it lets through light that is polarized parallel to the grain of the top plate. Now something interesting happens to the polarization direction of that light: it follows the twist of the liquid crystal, so it arrives at the bottom rotated 90°. The bottom Polaroid is oriented so as to let that rotated light through to the mirror. After reflection, the light makes the journey in reverse and comes out of the stack at the top. When that is happening, the background of the display looks light.

Only one thing left. How are the numbers made dark? To make dark in the display, it is only necessary to destroy the screw arrangement in the liquid crystal. An electric field will do that. It makes the molecules align parallel to the field. No more twist, so any light that passes one Polaroid will be stopped by the other. Electric field is applied via the coating on the plates, which is tin oxide, electrically conducting, very thin, and transparent.

A number is made up of seven segments (Fig. 3B). The segments are connected separately to the IC chip, to a part of it called the decoder-driver. That puts electric potential on the segments, in combinations to form any number 0 to 9. The number segments are a part of the tin oxide coating on the upper glass plate, but isolated electrically from the main area. The coating on the lower plate is uninterrupted and serves as the ground plane for the segments, in common. When the decoder-driver puts electric potential on a given segment, there is a field only below that segment,

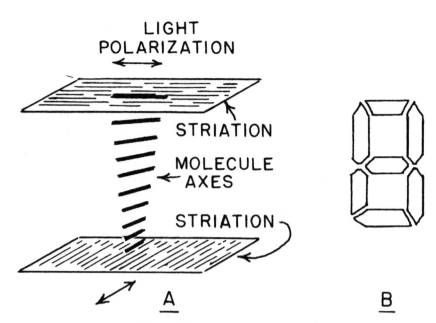

Fig. 3. **A**: How the axes of the liquid crystal molecules arrange themselves in a pattern that twists 90° from one plate to the other, so as to match the directions of striation on both plates. The light polarization follows the twist. **B**: The seven segments out of which all the numbers, 0 to 9, are formed. Note that all the cuts between segments are not the same. That is partly to accommodate the electrical connections, not shown.

destroying the twist, and making the segment look dark.

References

1. H.R. Crane, *Phys. Teach.* **31**, 501 (Nov. 1993).
2. Swiss Watch Technical Center, 1817 William Penn Way, Lancaster, PA 17601.
3. Digital watch "Hot Times," from "The 99-cent Store." Made in China.
4. H.R. Crane, *Phys. Teach.* **21**, 467 (Oct. 1983).

Mini-Motors Driven by Waves

Reading about a motor no bigger than a stack of four dimes and driven at 6000 revolutions per minute by the piezoelectric effect[1] brought me to quick attention. It was in a report[2] by Anita Flynn of the Artificial Intelligence Laboratory at MIT.[3] In reply to my questions, she sent several more reports and further information in a letter.

Piezo motors, also called ultrasonic motors or vibration motors, are quite new, little more than a decade old. They have been developed and put to practical application almost exclusively in Japan, and still are nearly unheard of in the U.S. industrial complex. But not so in the U.S. Patent Office, which lists more than 90 patents on the motors, all of Japanese authorship!

The piezo motor is simple in concept. Designs vary, but one will be enough to give all the ideas. The rotating part of the motor (the armature) is a metal ring that sits on, or is pressed down on, a stationary metal ring (the stator). A train of traveling waves circulates around in the stator (Fig. 1A). The waves are excited and kept going by the vibrations of piezoelectric crystals that are mounted on the underside of the stator. The armature is in contact only with the crests of the waves, so at first thought one might picture the armature being rotated at the rate of circulation of the wave crests. But wait: it's not that simple. The first surprising fact is that the armature rotates in the sense *opposite* to that of the circulating wave crests.

Of course the material of the stator ring does not travel; only the shape, or *profile*, travels. Its velocity is properly called phase velocity. If a mark were put on the ring, the mark would move up and down (transversely) with the passing of the crests and troughs of the waves. But not quite up and down. It's the "not quite" that makes the motor possible: the mark moves up and down in a thin loop, similar to an ellipse (Fig. 1B). So it has a small horizontal velocity when at the crest and the opposite when at the trough. As already said, the armature rests only on crests. It gets the benefit of the velocity in one direction and not the opposite, and that makes it go around.

At a physics brown-bag lunch we got to speculating as to how the motor idea might apply to waves in the ocean. We've seen swimmers out there (away from the effects of the beach) just bobbing up and down, going nowhere, as the crests and troughs go past them. What, we wondered, would happen if (thinking of the motor) they were on a raft that would span a number of waves so as to be supported on the crests only (Fig. 1C). Would the raft go? Probably. But which way? Against the profile, as in the piezo motor?

Colleague G.W. Ford undertook, with his students, to calculate how the motion at the crests of liquid waves might or might not differ from that at the crests in the elastic solid of the motor. For the elastic solid they could start with theory already in the books. Lord Rayleigh worked that out in 1885. For the liquid waves they found, to our surprise, that the velocity at the crests would be *in* the direction of travel of the profile of the waves. So the swimmers on the raft would go safely to shore, if the wave profile were going in, as it usually does. Thus a new form of ocean travel had been invented. Wonder if prehistoric seafarers knew that!

To come back to the motor, there are a few more details of the physics to be mentioned. First, the frequency of the driving piezo crystals is high: a typical figure is 40 kHz, and it may be up to 100. While the horizontal velocity imparted to the armature by the wave crest is small compared with that of its profile (phase velocity), the high frequency makes up for it. So the armature can rotate at quite a high rate: from hundreds of rpm up to thousands. The choice is wide, depending on the piezo frequency and other parameters.

Second, it is required that there be an integer number of waves in the closed ring of waves. The piezo crystals must be driven to vibrate at one of a series of

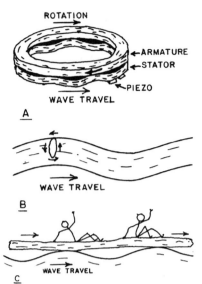

Fig. 1. A: Waves traveling counterclockwise in the stator, producing clockwise rotation of the armature. B: A part of the traveling wave in the stator, showing the approximately elliptical path of a particle of matter as one cycle of the wave passes by. C: A free ride on the crests of a traveling wave in liquid (the ocean). The raft is carried in the direction of the traveling waves. Referring to B, a particle of water moves clockwise around the ellipse, giving the raft forward velocity, but at much lower speed than that of the travel of the wave crests.

discreet frequencies. Typically there are 4 to 8 waves in the ring. The high frequency power for the piezo crystals of course comes from an external integrated circuit chip and associated electronic circuit.

Third is a requirement for making traveling waves, as against standing waves. A single piezo crystal applied to the ring would make standing waves. To make traveling waves, it is necessary to have a pair, or pairs, of crystals spaced a quarter wavelength apart, and vibrated at a quarter-cycle difference in phase. One pair would make the motor go, but in practice there is a series of pairs around the ring, so more power can be pumped in.

What do the Japanese use the motors for? I'll give two examples. In the Canon autofocus camera, the motor is used to turn the screw that moves the lens in and out. In the same camera it has also been applied to controlling the iris. The Toyota Company is applying the motors as actuators of mirrors, seats, and antennas in top-of-the-line cars. Of course the motor in all cases has to run both ways, but that is simple: done by changing the phases of the piezo vibrations.

In my teaching experience, I've found students have less experience with traveling waves than with standing waves. They miss some interesting points. To make traveling waves in, say a rope or other medium of finite length, reflection from the outer end has to be eliminated. That can be done with the rope by attaching the outer end to something that yields to just such a degree that all of the energy of the arriving waves is dissipated: none left to be sent back. That condition is called an impedance match, widely used wherever reflection is not wanted, such as in distributing television energy through coaxial cables.

Another demonstration might be done with a closed wave path. That would show the requirement for discreet frequencies, as well as of phased drivers.

Finally, there is a kind of wave not mentioned: surface tension waves. Would something resting on the crests go forward or backward? A good, but difficult, student project, if a wave tank is available.

References

1. There is a discussion of the piezoelectric effect and piezo crystals in a footnote to an earlier "How Things Work," *Phys. Teach.* **26**, 120 (1988).
2. The scoop on ultrasonic motors in Japan, Anita Flynn, *Scientific Information Bulletin of the Office of Naval Research*, Asian Office, April-May-June 1992, pp. 99–102.
3. Technology Square, Cambridge, MA 02139.

Physics from Dripping Water

There is in the Ann Arbor Hands-On Museum an exhibit we call the "Double Piddler." Separated water drops from two nozzles collide, to make interesting shapes (Fig. 1A). The water to the nozzles is supplied by a pump that gives pressure pulses at 60 per second (Hz), so 60 drops are delivered by each nozzle, per second, in phase. A strobe light giving 60 flashes per second illuminates the drops, so the whole array appears to be stationary in space. The viewer has no way of seeing that what appears to be a stationary drop is a different drop every 60th of a second. A knob for fine adjustment of the strobe frequency is provided, so it can be made slightly different from 60 Hz either way. That makes the chains of drops appear to move slowly downward, or upward and back into the nozzles. A little fluorescein is added to the water (which recirculates) to make a better show. The exhibit is a hit.

Exhibits similar to the one described are found in other "hands-on" type museums. All are derived from the pioneering experiments of Harold Edgerton[1] of MIT. A version of the demonstration was described in this journal by Joe Pizzo.[2]

I tried several times, with poor results, to photograph the collision of two drops. It had to be of a single collision, because the pattern varies from one collision to the next, 60 times a second. That led me to set up in my darkroom a simple, non-pulsed dripper, illuminated by continuous light, to learn more about photographing drops. What I found started me off in a new direction. Serendipity. The results might be classed as useless but interesting information.

I found that when illuminated and viewed at the angles shown in Fig. 1B, the falling drops appeared bright in certain intervals of height and were invisible in between. The lengths of the bright intervals and the separations between them increased downward as would be expected if the drops reflected the light at regular time intervals while falling. Since the pattern of light repeated with each drop, it was easy to make the measurements by holding a ruler near the path. From that and the law of falling bodies, the frequency of light flashes came out to be 23 Hz. The volume of the drops was measured, by counting them into a measuring cup, to be 0.11 cc, which is 0.3 cm in radius. The light evidently was sent back from the drops by reflection, not by refraction and internal reflection as in a rainbow, because there was no separation of colors.

What next? An hypothesis that can be tested. Suppose that when the drop breaks away from the nozzle, surface tension causes it to vibrate as in Fig. 1C. Then when its top side is illuminated, it would reflect light at the times when its top is near to being flat. So the test is to calculate the vibration frequency from data above, and see if the answer is about 23

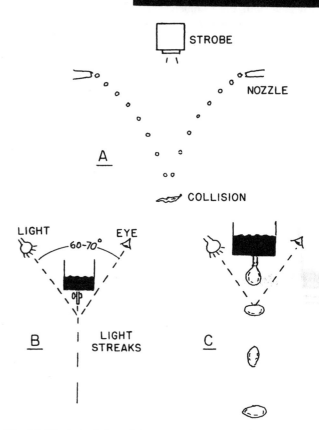

Fig. 1A. The two chains of drops in the Double Piddler, as they appear in the strobe light. All parts appear to be stationary, but the spatter of the collided drops changes 60 times a second. B. The simple dripper, the positions of the light source and the eye, and the streaks of brightness, as the drops fall. C. Imagined vibration of the falling drop, starting from the "teardrop" shape as it breaks away from the nozzle.

Hz. Only one additional number is needed: the surface tension of the water, 72 dynes/cm, from the handbook. I asked my colleague K.T. (Ted) Hecht, a nuclear theorist who is an expert with the nuclear liquid drop model, if he could calculate the frequency for this liquid drop. He did so readily, and got a frequency of 23.2 Hz,[3] which he gave with the remark that the agreement is too good. It is closer than should be expected, considering how the data were estimated, but it does confirm the hypothesis as to how the streaks of light are made.

One final test was made: to see if a change in surface tension would change the frequency. Laundry detergent was added to the water. As expected, the drops were smaller— 0.03 cc. The frequency of the light flashes went up to 28 Hz. Apparently the reduction in drop size increased the frequency more than the reduction in surface tension decreased it. The new surface tension was not measured, so a calculation was not possible.

Until this observation and one other came along I had thought that dripping water drops were spherical, only a little modified by moving through the air. The other observation is made from a window on the second floor above the ground, facing south. For as much as a day after a rain, water drips from the overhang of the roof, down past the window. With the Sun to the south, and looking downward from the window, the geometry is about that of Fig. 1B. Light from the drops is in intermittent streaks, but not as regular as shown in the figure. The drops may be both vibrating and tumbling.

References and Notes

1. Harold Edgerton, *Stopping Time* (Harry N. Abrams, Inc., New York, 1987), see p. 115.
2. Joe Pizzo, *Phys. Teach.* **25**, 512 (1987).
3. Professor Hecht found a formula given by A. Bohr, in *Dan. Mat. Fys. Medd.* **26** (14), 7–14 (1952): $\nu = \frac{1}{2\pi}\sqrt{\frac{8T}{\rho r^3}}$, where ν is frequency in Hz; T, surface tension in dynes/cm; ρ, density in g/cc; r, radius of the drop in cm. The formula is for vibration of the lowest frequency mode, and of small amplitude. The amplitude of the drops observed evidently is not small, which must introduce some error.